Fault Tolerance in Modern Engineering Systems

Edited by Mircea Ruba and Gabriel Chindris

Published in London, United Kingdom

Fault Tolerance in Modern Engineering Systems
http://dx.doi.org/10.5772/intechopen.1007234
Edited by Mircea Ruba and Gabriel Chindris

Contributors
Ahmed Dajani, Claudia Marţiş, Duo Zhang, Florin-Nicolae Jurca, Keiko Tsujioka, Mai Zheng, Mircea Ruba, Paula Şerban, Răzvan-Alexandru Inţe, Sebastian Ursache, Shahamat Shahzad Khan

First published in London, United Kingdom, 2026 by IntechOpen

IntechOpen is the global imprint of INTECHOPEN LIMITED, registered in England and Wales, registration number: 11086078, 167-169 Great Portland Street, London, W1W 5PF, United Kingdom

For EU product safety concerns: IN TECH d.o.o., Prolaz Marije Krucifikse Kozulić 3, 51000 Rijeka, Croatia, info@intechopen.com or visit our website at intechopen.com.

British Library Cataloguing-in-Publication Data
A catalogue record for this book is available from the British Library

Fault Tolerance in Modern Engineering Systems
Edited by Mircea Ruba and Gabriel Chindris
p. cm.
Print ISBN 978-1-83635-166-5
Online ISBN 978-1-83635-165-8
eBook (PDF) ISBN 978-1-83635-167-2

Meet the editors

Associate Professor at Technical University of Cluj-Napoca (TUCN), in the Department of Electrical Machines, Dr. Mircea Ruba is also a member of CAREESD, in charge of the Real-Time System Modelling and Testing Lab (RTsymLab). His research interests primarily revolve around electric vehicle (EV) drives, power electronics, and their real-time FPGA-based simulation and Hardware-in-the-Loop (HiL) testing. Having been involved in more than 22 research projects, Dr. Ruba has contributed significantly to the development of electrical motors, power inverters, controllers, and battery supply units for EVs. His research has led to the development of complete EV drivetrains designed for real-time simulators. He has also focused on Li-ion battery identification, simulation, and testing, resulting in a patented ultra-fast, efficient battery cell voltage-balancing unit. With a significant background in fault-tolerant drive solutions, converters, and controllers, Dr. Ruba's research has yielded patent-worthy innovations, including a novel fault-detection and compensation strategy and a modular fault-tolerant motor design. His patents have been recognized in 13 international exhibitions. Dr. Ruba's contributions include being an invited editor for the Applied Electromechanical Devices and Machines for Electric Mobility Solutions book, published by INTECH since 2020. The publications, patents, and projects in which he has been involved highlight his ability to design, analyze, and simulate in real time, and to conduct laboratory HIL testing of batteries and auxiliary components within an EV's propulsion unit.

An experienced researcher with more than 25 years of experience in teaching, designing, and testing embedded electronics, IEEE senior member, and holder of two registered patents, Dr. Gabriel Chindris is currently an Associate Professor with the Technical University of Cluj-Napoca (TUCN), Department of Applied Electronics, and is the head of ITEC Embedded Research Center of TUCN. His scientific interests are in advanced technologies for embedded systems, virtual prototyping, and Hardware-in-the-Loop (HiL) testing. He has been a researcher in over 30 international projects spanning automotive, consumer goods, testing and reliability, measurement equipment, and medical/assistive technologies, collaborating with partners from academia, industry, research centers, and start-ups. He has published over 90 conference papers, 2 ISI journal papers, 4 books, and 2 patents.

Contents

Preface

This book explores cutting-edge solutions for fault tolerance in modern engineering systems. Embracing a multidisciplinary and multi-domain perspective, it bridges areas such as electrical machines, robotics, avionics, autonomous vehicles, embedded systems, and artificial intelligence. The content is carefully curated to address both legacy strategies and emerging innovations that aim to enhance the resilience and operational continuity of complex systems.

The contributing authors present a balanced blend of theoretical insights and practical implementations, offering readers a deep understanding of how to design, analyze, and deploy fault-tolerant architectures. Special attention is given to techniques such as real-time error detection and correction, hardware redundancy, software-based recovery mechanisms, prognostics and health management (PHM), and adaptive reconfiguration.

A distinctive feature of this volume is its focus on both academic advancements and industrial applications. Case studies from aerospace systems, smart grids, intelligent transportation, and mission-critical automation illustrate how fault tolerance is implemented in real-world scenarios. From laboratory prototypes to production-ready technologies, the methods discussed are validated through experimental data, simulation models, and analytical evaluation.

The book is particularly valuable for engineers, researchers, and graduate students involved in the development of high-reliability systems. It provides actionable frameworks for fault detection, diagnosis, and recovery, alongside tools for reliability modeling and system-level risk assessment. Furthermore, chapters addressing AI-assisted fault management, digital twins for resilience simulation, and cyber-physical system protection reflect the evolving role of machine learning and data-driven techniques in fault-tolerant design.

Overall, this work serves as both a foundational reference and a practical guide for anyone engaged in building systems that must operate safely, reliably, and continuously, even in the presence of faults.

Mircea Ruba and Gabriel Chindris
Associate Professor,
Technical University of Cluj Napoca,
Romania

Chapter 1

Modular and Fault-Tolerant Topologies of Outer Rotor Synchronous Reluctance Machines for Small Electric Propulsion Systems

Răzvan-Alexandru Inţe and Florin-Nicolae Jurca

Abstract

This chapter begins with a brief introduction to alternative transportation solutions and current trends. It discusses the types of electric motors used in low-power propulsion systems along with their advantages and disadvantages. Continuing with the theme of modularity, the chapter presents the concepts of modular and modular-segmented electric machines and compares them to conventional ones. These rotor designs offer fault tolerance and easy disassembly capabilities for electric motors. The trend of using electric motors that do not rely on rear earth materials is well known. Therefore, we chose a synchronous reluctance machine (SynRM) with an outer rotor for a detailed analysis. The SynRM is analyzed in classic, modular, and modular-segmented configurations using finite element analysis (FEA) with FLUX 2D and MATLAB. The chapter also analyzes the slots/poles combination and flux barrier shape of electric machines. Finally, the results obtained from FEA simulations are compared with the built prototype.

Keywords: electric machine design, outer rotor, fault tolerance, modular construction, electric propulsion, synchronous reluctance machine, facile disassembly and maintenance

1. Introduction

With rapid economic growth and the acceleration of urbanization, urban populations have increased quickly, leading to intolerable traffic congestion in many cities [1]. Urban transportation is particularly challenging in densely populated areas, where traffic jams occur daily at peak hours. Several key aspects must be considered when evaluating urban mobility, including commute time, air pollution, financial factors, and versatility. Since 1750, human activities have been the primary driver of increasing greenhouse gas (GHG) concentrations. GHG levels have continued to rise since 2011, reaching an annual average in 2019 of 410 ppm for carbon dioxide (CO_2),

1866 ppb for methane (CH_4), and 332 ppb for nitrous oxide (N_2O) [1]. These gases contribute to global warming and pose health risks due to their toxic effects when present in the atmosphere [1]. In Europe, environmental legislation has undergone changes, with low-emission vehicles permitted in certain cities while GHG emission limits have been imposed on manufacturers of cars, vans, and trucks by the European Union [2, 3]. Currently, between 10% and 30% of global GHG emissions are attributable to the transportation industry [4]. Because of legal requirements, nearly every automaker now offers at least one electric vehicle (EV) or plans to do so [5]. The European Commission postponed the adoption of the next Euro 7/Euro VII emission standard until April 5, 2022, but according to the European Automobile Manufacturers' Association (ACEA), light-duty vehicles should start using Euro 7 on September 1, 2025 [6], while heavy-duty vehicles should start using it on December 31, 2025 [7].

In the last decade, urban mobility has evolved significantly in big cities. There are several aspects to consider when choosing the right vehicle for commuting, including vehicle cost, transportation costs, travel distance, vehicle range, and taxes and insurance. EVs appear to be the best option for reducing GHG emissions from urban transportation in most European countries. However, given the short travel distances in urban areas, many people prefer to travel by bicycle. A variety of individual electric transportation options are available, such as electric bikes (e-bikes), motorbikes, trikes, hoverboards (with or without handlebars), skateboards, unicycles (eWheels), and rideables, all of which are currently available on the market [8]. Recently, e-bikes have become one of the most popular alternative transportation options because they require less mechanical input from riders and provide assistance when cycling uphill. This makes e-bikes particularly beneficial for elderly people.

Trends indicate that e-bike adoption in the Netherlands varies according to the age and demographics of users. For instance, older people choose e-bikes when traditional riding becomes difficult, whereas younger users transition from cars to e-bikes. These tendencies are expected to change as e-bikes become more popular with various populations. These new trends point to a generational shift in mobility preferences, which merits more research on the effects of e-bikes on various generations and demographics [9, 10].

Compared to a traditional bicycle, an e-bike integrates several key components that enhance its functionality but also require specialized maintenance. These include an electric motor, which assists pedaling; a rechargeable battery, which powers the system; a controller, which regulates power distribution; and various sensors, such as torque and cadence sensors, that optimize performance. Additionally, wiring and electronic connections must be inspected regularly to prevent faults, making e-bike maintenance more complex than that of a conventional bicycle [11]. In addition to standard break, tire, chain, and gear adjustments, e-bikes require additional maintenance to their electric propulsion system, including the battery, power converter, electric motor, connection interfaces, and control unit. This maintenance should be done in authorized shops by trained and certified personnel.

2. Electric motors

The first step in designing an electrical machine is to establish the system requirements. These requirements may be dictated by the application, client, or legislation. One of the most common requirements directly influencing electric motor design is the maximum output power permitted by law, particularly in the case of e-bikes.

2.1 Electric machine requirements

The electric machine requirements vary depending on the specific EV application and are influenced by the following parameters: available volume, power density, torque density, vehicle speed, energy supply, and gearbox configuration. This section focuses on the electric machine requirements for e-bikes; however, the following design principles can also be applied to other domains.

While all small electric propulsion vehicles are covered by European Regulation 168/2013, two distinct categories are of particular interest: L1e-A for powered cycles and L1e-B for mopeds, where certain manufacturing requirements, including maximum speed and power, are enforced. All e-bikes are included in the L1e-A, and the maximum permitted speed is 25 km/h. Additionally, the maximum continuous rated power should not be greater than 1000 W. The L1e-B is designed for mopeds, with a top speed of 45 km/h and a maximum continuous rated power of 1000–4000 W [12].

Another important aspect of motor design is its placement within the propulsion system. There are three primary locations where the motor can be mounted (**Figure 1**) [13].

2.2 Electric machine types

There are several types of electric machines available on the market. Choosing the appropriate machine for a given application can be challenging. Multiple factors must be considered, including mechanical, electrical, and thermal requirements, machine volume, and overall cost. **Table 1** compares various electric machine types based on key characteristics.

The first electric machine used in propulsion applications was the DC motor, favored for its cost, technological maturity, and simple control technique. Even so, DC machines have major disadvantages, including periodic maintenance due to mechanical commutation, low power density, and poor efficiency. Induction machines (IMs) have also been used in electric propulsion, particularly in railway and EV applications due to their low cost and high reliability. However, despite these benefits, IMs are unable to meet propulsion requirements due to their low efficiency, low power factor, and low power density, which becomes more noticeable in low-power applications.

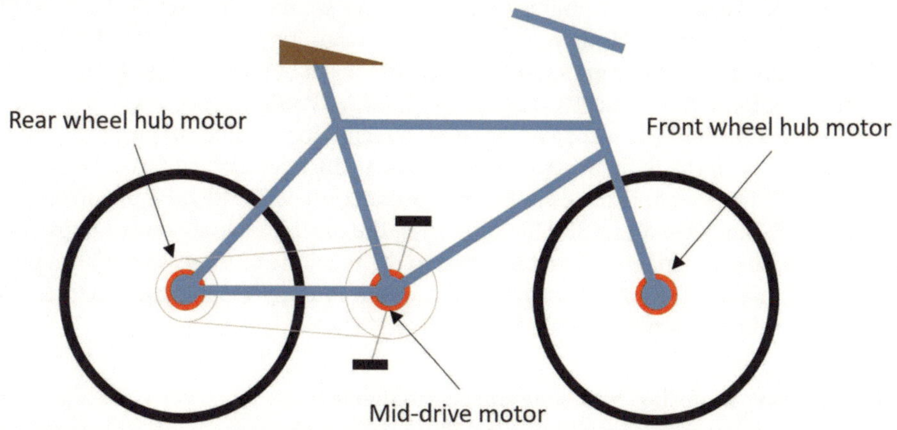

Figure 1.
Electric motor placement possibilities for an e-bike.

	DC	BLDC	SRM	IM	PMSM	SynRM	PM-SynRM
Power density	1	5	4	3	5	3	4
Efficiency	2	5	3	3	5	3	4
Power factor	3	4	2	3	4	3	4
Robustness	2	3	5	5	3	5	3
Costs	4	2	5	5	2	5	4
Thermal behavior	2	2	3	4	2	4	3
Control	5	3	2	3	4	3	3
Legend	1 – very bad, 2 – bad, 3 – neutral, 4 – good, 5 – very good						

Table 1.
Comparison of electric machines technologies [14].

The classic DC motor was eventually replaced by the brushless DC (BLDC) machine due to advancements in electronics and materials science. The BLDC machine is currently one of the most popular electric machines for low-power propulsion, such as e-bikes and mopeds, and is utilized by major manufacturers like Bosch, Yamaha, Brose, Shimano, Fauza, Specialized, and others because of its high power density and efficiency [15]. However, BLDC machines use rare earth magnets, which increase costs compared to other motor types.

For high-power applications, the permanent magnet synchronous motor (PMSM) is preferred due to its reduced torque ripple compared with BLDC machines. PMSMs offer high power density and efficiency, making them ideal for propulsion applications. However, their reliance on rare earth permanent magnets is a disadvantage, and high-speed operation can present back-EMF challenges.

Switched reluctance machines (SRMs) are a cost-efficient compromise for electric propulsion applications since they can achieve a high power density and are reasonably priced and efficient [16]. Although the machine's sturdy and straightforward design makes it dependable under stable mechanical and thermal conditions, the double-salient structure causes high torque ripple, high vibration, and noise [17]. A non-standard inverter and a custom control method are required to operate SRMs.

Synchronous reluctance machines (SynRMs) offer significant potential for electric propulsion because of their numerous advantages, including their compatibility with standard AC inverters, lower material costs because permanent magnets are not needed, and absence of back-EMF issues. Compared to SRMS, SynRMs cause less vibration and noise. Historically, SynRMs were avoided due to technological limitations in the rotor's construction, which made them structurally weak against centrifugal forces. However, technological advancements have led to mass production of SynRMs by major corporations like ABB, Siemens, and KSB [18]. Although the machine's low power factor is a drawback, permanent magnets can be added to the machine to increase both the power factor and efficiency. Doing so transforms the motor into a permanent magnet-assisted synchronous reluctance machine (PMA-SynRM) [18].

2.3 Modularity, modular, and segmented concepts

The design of electric machines for propulsion applications is more complex than simply analyzing the electromagnetic behavior of the machine. From a design perspective, a multi-physics approach is required, incorporating electromagnetic, mechanical, thermal, noise, and vibration analyses. The modular design of electric machines is an innovative approach that enhances manufacturing, performance, and maintenance. Instead of constructing a machine as a single, monolithic unit, modular designs use standardized and interchangeable components. This allows for greater flexibility in production and customization for various applications.

Manufacturing and maintenance are two important factors in electric machine design. From a maintenance perspective, modular designs simplify repairs and replacements. Instead of disassembling an entire motor, faulty modules can be individually swapped, minimizing downtime. This is particularly beneficial in industries where continuous operation is critical.

Therefore, a modular electric machine offers several advantages:

- Easy manufacturing and assembly
- Simplified and low-cost maintenance
- Reduction in production and operational costs
- Design for disassembly
- Sustainability and recyclability
- Increased efficiency

Despite these advantages, modular machine designs come with some challenges, including increased complexity in design, potential performance reductions in specific regimes, and mechanical, structural, or vibration problems.

There are three primary concepts in electric machine construction: modularity, modular, and segmented designs. The modularity concept refers to an electric machine designed at a rated power of 10 kW, with the capability to scale up to 30 kW by stacking three machines together (**Figure 2a**). This approach provides high fault tolerance since multiple modules can function independently. In the event of a winding failure in one module, the machine can still operate at a lower output power. This concept is widely used in various electric traction applications.

For SynRM, a modular rotator design offers higher torque density compared with conventional construction while reducing iron losses in the rotor core. The second approach involves modular construction in rotor, stator, or both. **Figure 2b** shows a modular rotor concept in reverse construction, which provides several advantages, including simpler manufacturing, easy assembly, and higher saliency ratio compared to a traditional rotor [19]. In addition, modular construction also improves maintenance and reduces costs, as individual components can be replaced, extending the machine's lifespan.

The third approach is the modular segmented rotor (**Figure 2c**), which offers a high saliency ratio. However, production complexity increases due to the need for resin or spacers between rotor core components. For secure assembly, rotor parts must be glued together, making individual segment replacement challenging in case of failure. In this case, the best solution overall is the modular rotor.

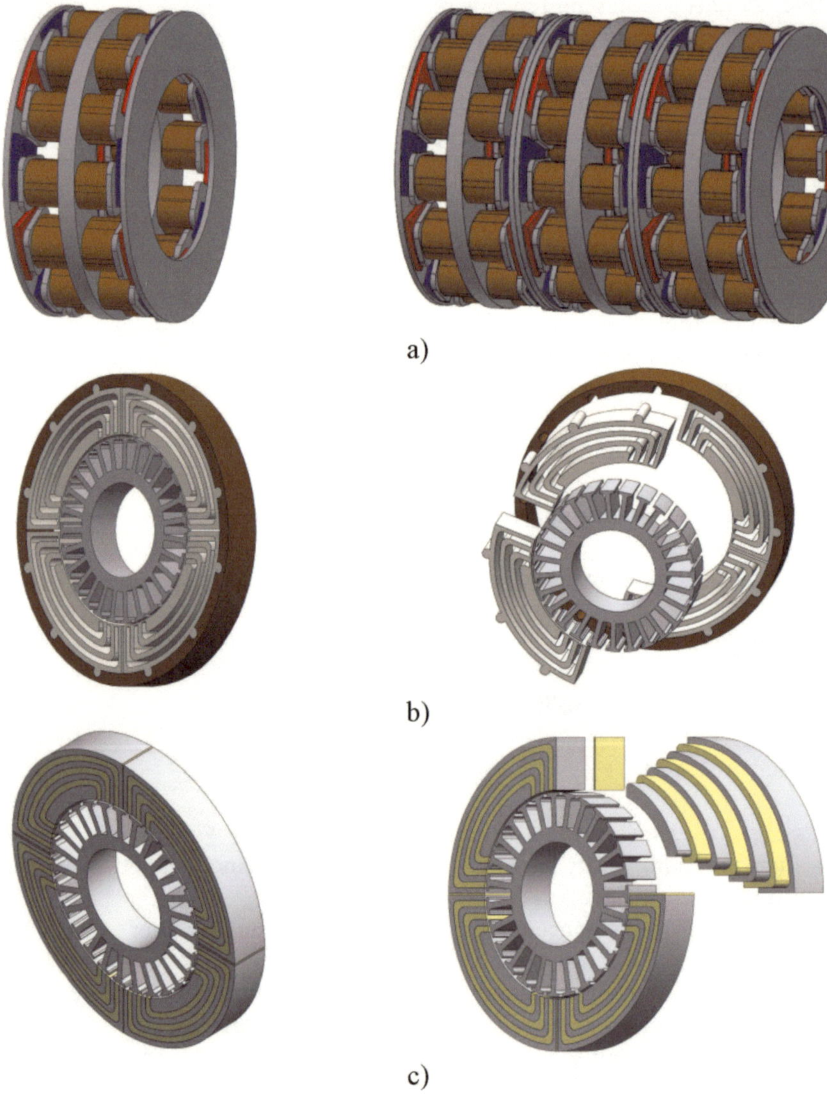

Figure 2.
Electric machine construction concepts (a) modularity, (b) modular, and (c) segmented.

2.4 Synchronous reluctance machine

The rotor design of a SynRM has a major impact on its performance. Unlike electric machines, SynRMs lack both permanent magnets and rotor windings. Consequently, the dynamic rotor reference frame equations for d-q axis voltages are given as follows:

$$V_d = R_s i_d + L_d \frac{di_d}{dt} - \omega_e L_q i_q \tag{1}$$

$$V_q = R_s i_q + L_q \frac{di_q}{dt} + \omega_e L_d i_d \tag{2}$$

The electromagnetic torque is expressed as:

$$T_{dq} = \frac{3}{2}p\ (\Psi_d I_q - \Psi_q I_d) \tag{3}$$

where Ψd, Ψq, Id, and Iq represent the d-q components of flux linkage and currents, and p is the number of pole pairs in the rotor. Flux linkage equations can be expressed as:

$$\Psi_d = L_d I_d \tag{4}$$

$$\Psi_q = L_q I_q \tag{5}$$

So, the d-q axis inductance can be defined as:

$$L_d = \frac{\psi_d}{id} \tag{6}$$

$$L_q = \frac{\psi_q}{iq} \tag{7}$$

Therefore, according to Eqs. (3)–(5), the corresponding electromagnetic torque Te can be written as:

$$T_e = \frac{3}{2}p\ (L_d - L_q) I_d I_q \tag{8}$$

The electromagnetic torque is influenced by three key factors: the number of pole pairs, the difference between the d-q axis inductances, and the d-q axis supply currents. The number of pole pairs can be optimized during the design process and kept as high as possible to maximize torque. However, a well-known drawback of this type of machine is its low power factor. Both the power factor (cosφ) and electromagnetic torque depend on the d-q axis inductance. Specifically, cosφ is directly affected by the saliency ratio, as shown in Eqs. (9) and (10).

$$\xi = \frac{\mathrm{L_d}}{\mathrm{L_q}} \tag{9}$$

$$\cos\varphi = \frac{(\xi - 1)}{(\xi + 1)} \tag{10}$$

The d-q axis inductances Ld and Lq were calculated using Eqs. (6) and (7). Once these values were determined, the next step was to compute the saliency ratio (Eq. 9) and the power factor (Eq. 10). Using these equations, the variation of the power factor as a function of the saliency ratio is illustrated in **Figure 3**. The SynRM achieves a power factor of 0.6 when the saliency ratio exceeds 4, and reaches 0.8 when the saliency ratio surpasses 10, which is challenging to achieve. It can be observed that attaining a higher power factor requires a significantly high saliency ratio.

The in-wheel traction motor for e-bikes is one of the simplest and most cost-effective solutions available on the market. Several companies, including Bafang, Shengy, Dapu, MXUS, Mahle, and Zehus, offer conversion kits that can transform a regular bicycle into an e-bike. The characteristics of the traction motor vary depending on the type of e-bike. The electric machine is integrated into the wheel rim

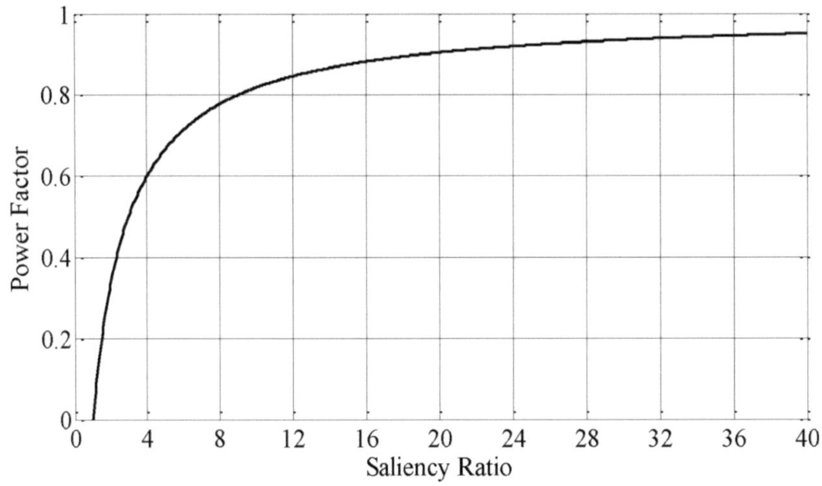

Figure 3.
Power factor variation in the function of saliency ratio [19].

and can be used with different wheel diameters, ranging from 13 to 29 inches. Most motors on the market are either PMSM or BLDC machines; however, the proposed machine is an outer rotor synchronous reluctance machine (ORSynRM).

2.5 Design of ORSynRM

The ORSynRM encounters identical design issues as the inner rotor (IR) SynRM. Due to the non-linear magnetic field characteristics of SynRMs analytical equation techniques are insufficient for design validation. Consequently, a numerical analysis utilizing the Finite Element Method (FEM) is essential. For light-vehicle applications, specific electrical and geometrical parameters are imposed. Their values are listed in **Table 2**.

2.5.1 Slot/poles combinations

The design process begins with a topological analysis to ascertain the optimal slots/poles configuration. The initial focus is on structures with a high number of poles and slots. These structures are analyzed from an electromagnetic perspective, focusing on produced torque, torque ripple content, saliency ratio, and power factor. Upon establishing the OR-SynRM topology, the subsequent stage involves enhancing the rotor design to augment the saliency ratio and the machine's power factor.

In BLDC and PMSM topologies, a higher number of slots and poles is typically used. Thus, the ORSSynRM initially followed a similar approach in its slots/poles combination. However, **Table 3** shows that having a larger number of poles does not guarantee a higher torque density, which is necessary for propulsion applications. Instead, the average electromagnetic torque increases when the number of rotor poles decreases. In addition, the right combination of stator slots helps maintain low torque ripple content. According to the results presented, the 4-pole, 27-slot configuration has the best overall performance for the ORSynRM.

Parameter	Value	Unit
Rotor outer radius	0.115	[m]
Rotor inner radius	0.0898	[m]
Stator outer radius	0.089	[m]
Stator inner Radius	0.062	[m]
Air gap heigh	0.006	[m]
Iron core length	0.05	[m]
No. of phases	3	—
Fill factor	~0.5	—
Phase voltage	24	[V]
Output Power	340	[W]

Table 2.
Geometrical and electrical constraints of the designed ORSynRM.

No. of slots	No. of poles	Mean torque [N·m]	Torque ripple [%]	Saliency ratio	Power factor
57	20	2	14.13	1.97	0.327
	16	4.79	24.16	2.56	0.438
	10	7.53	4.65	2.59	0.443
54	20	0.82	6.55	3.41	0.546
	16	3.72	9.75	2.53	0.433
	10	6.81	9.35	2.87	0.484
51	20	1.75	8.78	2.33	0.399
	16	1.54	3.7	1.35	0.15
	10	4.14	6.82	2.28	0.39
36	10	4.97	13.5	2.06	0.347
	8	6.55	22.4	2.59	0.442
	6	7.6	62.52	2.75	0.466
	4	5.6	76.9	1.82	0.291
27	10	1.9	20.7	1.61	0.236
	8	7.8	16.3	2.4	0.413
	6	7.5	8.2	2.43	0.416
	4	**7.9**	**8.8**	**2.51**	**0.441**

Table 3.
Performance comparisons for different slots/poles configuration of the ORSynRM.

2.5.2 Flux barrier shapes and number analysis

Based on the topology with the highest performance (4 poles, 27 slots) identified previously, an analysis was conducted to determine how rotor flux barrier shapes influence the key performance parameters of the SynRM. Four flux barrier shapes were analyzed: round, rectangular, trapezoidal, and a round-rectangular combination. The simulations were run using Flux 2D coupled with MATLAB to automatically modify rotor geometry within the imposed geometrical constraints and compute the saliency ratio, followed by a transient magnetic simulations for the best-performing rotor to compute electromagnetic torque and torque ripple content. For each flux barrier shape, three rotor configurations were studied with one, two, and three flux barriers per pole. However, the maximum number of flux barriers per pole was limited due to outer diameter constraints.

Rotor configurations featuring three flux barriers per pole pose a risk due to thin ribs (less than 3 mm thick), compromising structural integrity. As a result, only designs with one or two flux barriers per pole are suitable for this application. Among the evaluated structures, the round and trapezoidal rotors are the heaviest, while the rectangular and round-rectangular rotors are lighter. Although the configuration with three round-rectangular flux barriers achieves the highest power factor, it also exhibits significant torque ripple. Therefore, the rotor with rectangular flux barriers was selected for further study due to its reduced weight, structural robustness, and high electromagnetic torque generation.

Figure 4 illustrates the flux barrier shapes, showing the configuration with two flux barriers per pole. At the same time the magnetic flux density can be also visualized. To ensure a fair comparison over all designs, the stator and winding characteristics were kept constant in all machines.

According to **Table 4**, the lightest rotor is the round-rectangular flux barrier shape with three flux barriers per pole. However, the thin 2 mm ribs between flux barriers compromise structural integrity. Given that the ORSynRM is intended for e-bike propulsion, ensuring the structural integrity of the machine is crucial.

2.5.3 Fault-tolerant designs

Fault-tolerant machines and drives are highly desirable for electric traction and aeronautical applications, as they ensure reliable performance even under fault conditions. In safety-critical applications, electric drives must exhibit fault tolerance to maintain high availability [20]. One approach to achieving fault tolerance is the integration of two or more machine modules, either in parallel or in series [21, 22]. Various types of electric machines, such as permanent magnet synchronous reluctance machines (PMSM), switched reluctance machines (SRM), and flux switching machines (FSM), have been designed to operate under fault-tolerant conditions (**Figure 5**) [23–25].

The modular and segmented SynRM rotors were designed and simulated under the same conditions as the previous design in Section 2.5.2. Comparing the results from the classic rotor design (**Table 4**) with the modular and segmented rotor structures (**Table 5**), it is evident that both the modular and segmented designs exhibit higher torque and torque ripple content. While the classic structures have lower torque ripple, their torque output is reduced by 6.02% for the trapezoidal flux barrier configuration and 14.07% for the classic rectangular configuration. However, the modular rotor structure offers a significant advantage, increasing the power factor by more than 20% compared to the classic design.

The segmented rotor structure produces lower torque and torque ripple than the modular topology. In terms of manufacturability, the classic rotor design involves only the punching and stacking of steel sheets, but in case of structural failure, the entire rotor must be replaced. In contrast, the modular design allows for the independent punching

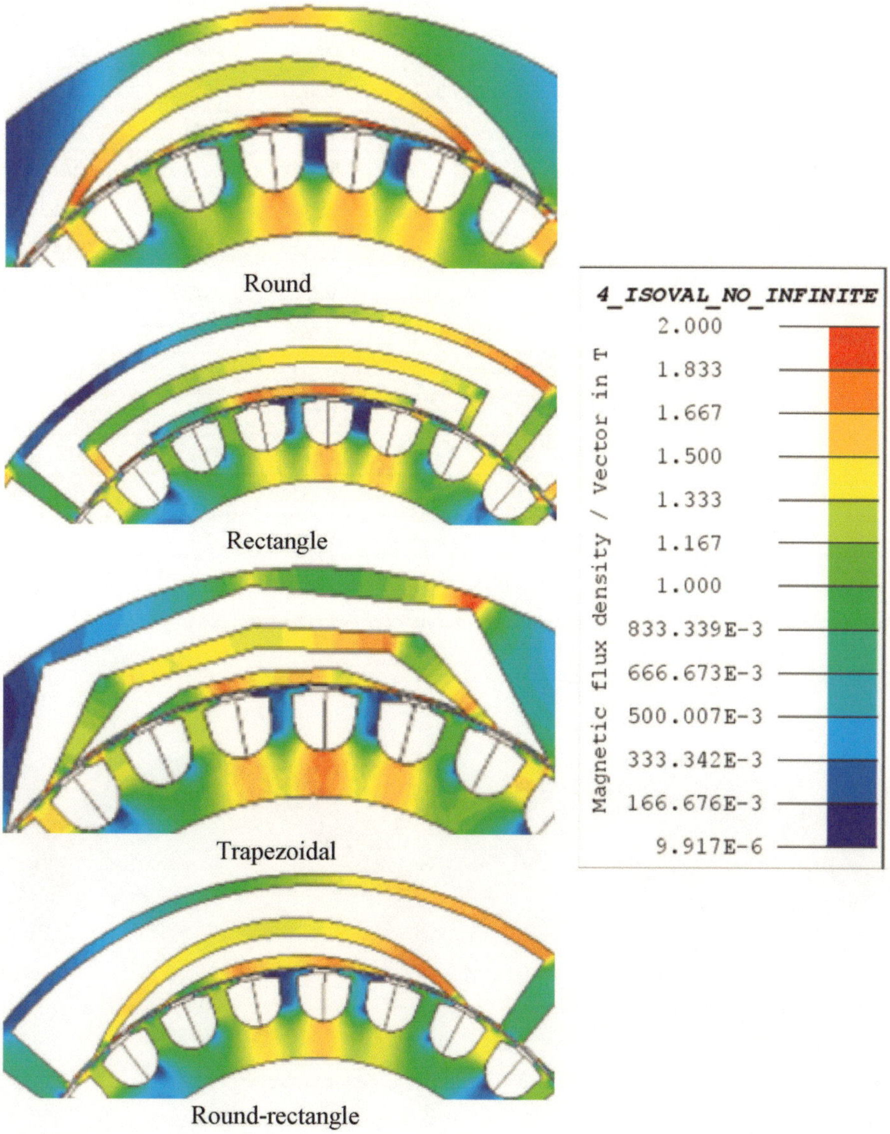

Figure 4.
Flux barrier shapes and magnetic flux density distribution.

and stacking of each module, enabling the replacement of individual poles. While this increases initial production complexity, it enhances maintainability.

For the segmented topology, mechanical failures can be addressed by replacing only the damaged rib within a pole. However, this approach significantly increases manufacturing complexity, as each rib must be individually punched, followed by rotor assembly in a plastic frame or epoxy resin to ensure structural integrity and fixation. Additionally, the fixation material adds extra mass to the rotor. Given that the segmented rotor does not offer superior electromagnetic performance compared to the modular design and involves significantly higher manufacturing complexity, it is not suitable for the proposed application. Therefore, the modular ORSynRM will be further analyzed in detail from electromagnetic, thermal, and mechanical perspectives.

Rotor Type	No. of barriers	Rotor Weight [kg]	Saliency ratio	Power factor	Torque [Nm]	Torque ripple [%]
Round	1	3.52	3.72	0.577	5.9	26.1
	2	3.83	2.43	0.417	8.04	21.5
	3	3.62	3.66	0.571	7.23	16
Rectangle	1	3.4	3	0.5	6	37.3
	2	2.45	3.29	0.534	7.75	23.7
	3	**2.36**	**3.32**	**0.537**	**8.18**	**20.1**
Trapezoidal	1	3.76	3.65	0.569	6.94	15.9
	2	3.72	3.28	0.532	8.46	15.1
	3	3.64	3.27	0.532	8.46	10.9
Round-rectangle	1	2.21	3.72	0.576	7.1	13.1
	2	2.42	3.78	0.582	8	19.9
	3	2.09	3.84	0.587	8.1	15.7

Table 4.
Performance results for different flux barrier shapes and number of barriers per pole.

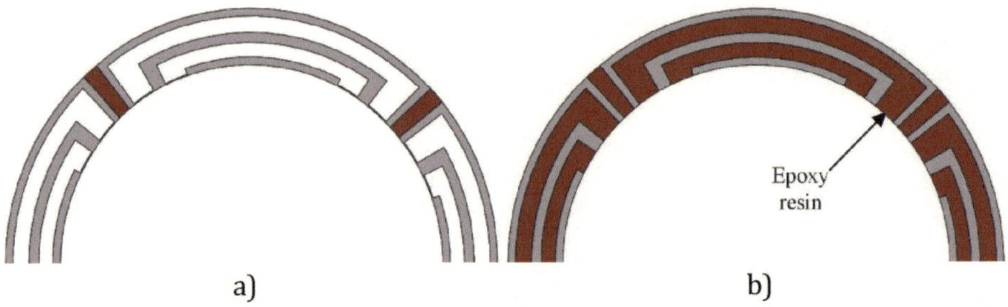

Figure 5.
Modular and segmented rotor structure. (a) Modular rotor. (b) Segmented rotor [19].

Rotor Type	No. of barriers	Rotor Weight [kg]	Saliency ratio	Power factor	Torque [Nm]	Torque ripple [%]
Modular	1	2.51	4.31	0.623	8.76	36.39
	2	2.72	5.11	0.672	9.02	20.02
	3	2.94	4.85	0.658	8.87	16.42
Segmented	1	2.37	4.94	0.663	8.17	33.1
	2	2.51	5.14	0.680	8.42	19.02
	3	2.55	5.02	0.668	8.62	29.93

Table 5.
Performances of modular and segmented ORSynRM.

3. Prototyping and validation

The modular ORSynRM was built according to imposed geometrical, electrical, and mechanical requirements. The motor construction company was only involved in designing the case to meet specific mechanical constraints. The following images show the rotor's magnetic core and the machine's casting. To simplify fastening, the manufacturer proposed that each rotor pole should have two fastening points, which are directly integrated into the casting through fastening channels. Since the machine operates at a low speed of 360 RPM, there is no need to use resin or any other material to secure the rotor poles in place. The rotor core is made from electrotechnical steel sheets (M270_35A) using a laser cutting process. Further the stator winding was placed in the stator core manually, and the rotor and stator were mounted together (**Figure 6**).

To validate the machine design, the electrical parameters of the machine are measured and compared with the calculated ones. Several parameters that are necessary for efficient control include phase resistance, phase inductance, and d-q axes inductances.

The first parameter measured is phase resistance, which can be determined using two methods. The first method involves supplying the ORSynRM with 1 ampere from a DC source and allowing the voltage to stabilize. Using Ohm's law, the total resistance is calculated as 2R = U/I, where U is the voltage amplitude, and I is the current flowing through the winding. Since the current and voltage pass through both phase A and

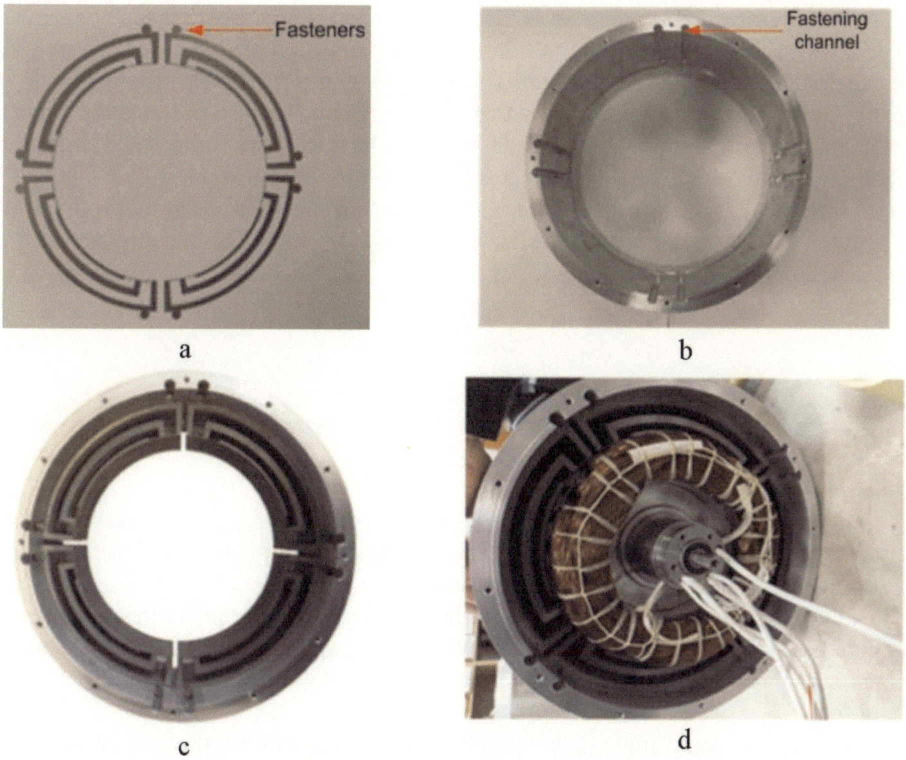

Figure 6.
(a) Modular rotor poles steel sheet, (b) aluminum housing with fastening channel, (c) rotor modules mount into the case, and (d) the stator assembly together [19].

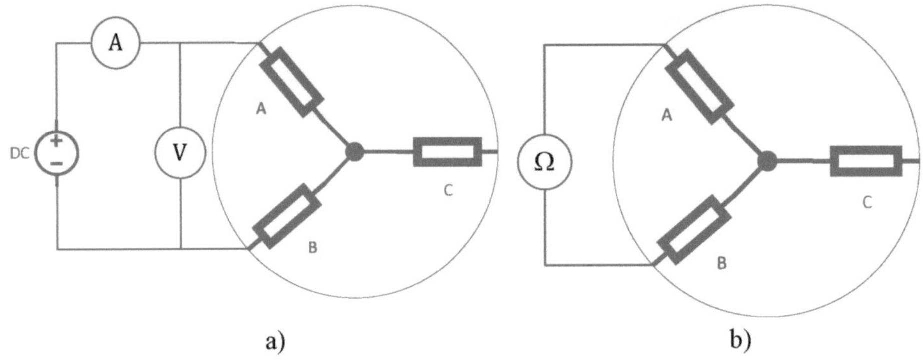

Figure 7.
Phase resistance measurement circuits. (a) DC source circuit. (b) Multimeter circuit.

phase B, the resulting resistance must be divided by two to obtain the phase resistance (**Figure 7a**). The second method involves directly measuring the resistance using a multimeter (**Figure 7b**). Like the first method, the measured resistance must be divided by two to determine the resistance of a single phase.

After applying both measurement methods, the phase resistance was found to be 0.162 Ω, which is lower than the previously calculated theoretical value. One possible explanation for this discrepancy is that the end winding turned out to be shorter than expected, likely due to the motor being wound manually.

The second measured parameter is phase inductance (Lph), which is easily determined using an LCR meter. Measurements are taken between phases A-B, B-C, and C-A to check for any variations. The differences between phases at the same frequency do not exceed 3%, indicating that the measurements are accurate. **Table 6** presents the results.

The mathematical model of the SynRM shows how d-q axis inductances influence machine performance. The next step in validation is to measure the d-q axis inductances, which can be determined using several methods. The simplest method is to use the previous measurements from **Table 6** and Eqs. (11) and (12) to compute the d-q axis inductances (**Figure 8**).

$$L_d = \frac{L_{ABmax}}{2} \tag{11}$$

$$L_q = \frac{L_{ABmin}}{2} \tag{12}$$

Table 7 presents the results.

Since the LCR meter operates within a limited voltage range, inductance measurements were obtained only for a small current range, which presents a significant drawback [27], as it does not account for saturation effects.

To obtain a clearer view of the d-q axis inductances, an alternative method must be employed [28]. This approach requires additional equipment, including a DC source, current sensor, voltage sensor, dSpace MicroLabBox, and a test bench for mounting the motor.

By varying the rotor angle θ_{el} it becomes possible to measure the inductance along the d and q axes. In this setup, L represents the total inductance for the series-parallel connection of the stator winding, and the corresponding L_d and L_q values are determined as follows:

Parameter	Frequency	Min value	Max value	Unit
LA-B	100 Hz	0.00723	0.0209	H
	120 Hz	0.00714	0.0204	H
	1 kHz	0.00688	0.0192	H
	10 kHz	0.00613	0.01356	H
LB-C	100 Hz	0.00708	0.0207	H
	120 Hz	0.00701	0.0201	H
	1 kHz	0.00676	0.0187	H
	10 kHz	0.00604	0.01343	H
LC-A	100 Hz	0.00711	0.0211	H
	120 Hz	0.00704	0.0205	H
	1 kHz	0.00674	0.0188	H
	10 kHz	0.00601	0.01351	H

Table 6.
Phase inductance results from LCR meter measurement.

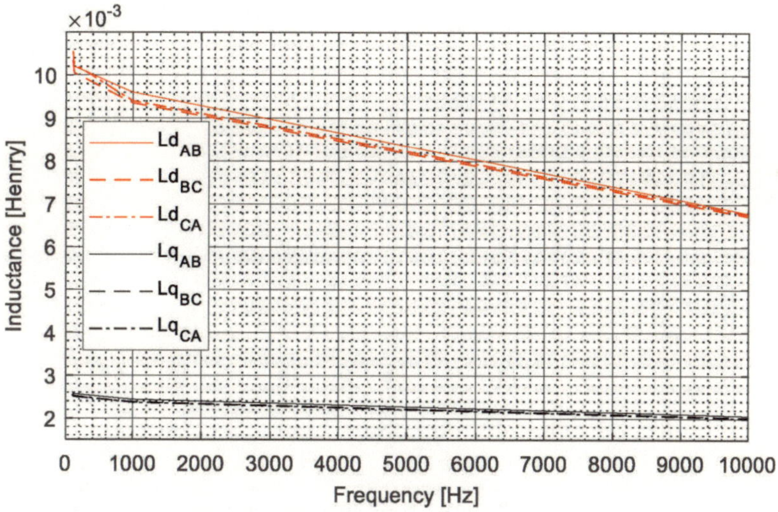

Figure 8.
The d-q axis inductance measurement in function of frequency [26].

$$L_d = \frac{2}{3}L, \ (\theta_{el} = 0^\circ) \tag{13}$$

$$L_q = \frac{2}{3}L, \ (\theta_{el} = 90^\circ) \tag{14}$$

When the rotor is aligned with phase A, $\theta_{el} = 0°$, and when the rotor is between phases B and C $\theta_{el} = 90°$. The rotor is mechanically locked in each position, and the current response in a first-order RL circuit is given by:

Parameter	Frequency	Lq	Ld	Unit
LA-B	100 Hz	0.003615	0.01045	H
	120 Hz	0.00357	0.0102	H
	1 kHz	0.00344	0.0096	H
	10 kHz	0.003065	0.00678	H
LB-C	100 Hz	0.00354	0.01035	H
	120 Hz	0.003505	0.01005	H
	1 kHz	0.00338	0.00935	H
	10 kHz	0.00302	0.006715	H
LC-A	100 Hz	0.003555	0.01055	H
	120 Hz	0.00352	0.01025	H
	1 kHz	0.00337	0.0094	H
	10 kHz	0.003005	0.006755	H

Table 7.
d-q axis inductances according to the LCR measurement method.

$$i_d = \frac{U}{R_s}\left(1 - e^{\frac{t}{\tau}}\right) \tag{15}$$

Where τ represent the circuit time constant:

$$\tau = \frac{L}{R_s} \tag{16}$$

Therefore, since τ is known the L_d, L_q inductances can be determined using the following equations:

$$L_d = \frac{2}{3}\tau R_s \tag{17}$$

$$L_q = \frac{2}{3}\tau R_s \tag{18}$$

The time constant τ is defined as the time interval between the voltage step initiation and the moment the current through the phase reaches 63.2% of the maximum value.

To determine the d-axis inductance L_d, the circuit shown in **Figure 9** was set up. The first step is to connect the positive terminal of the DC source to phase A, while the negative terminal is connected to phases B and C. A low voltage is applied to limit the current, ensuring that the rotor aligns with phase A. Next, the rotor shaft must be mechanically locked. Finally, a negative step voltage is applied to phase A, while phases B and C are connected to the positive terminal, as illustrated in **Figure 9b**.

In this case, several tests were conducted with the motor supplied with a DC voltages ranging from 0.5 to 7 V, in increments of 0.5 V. At the maximum voltage limit of 7 V, the current reaches the maximum value of 26.98A. The results from the first step are shown in **Figure 10**.

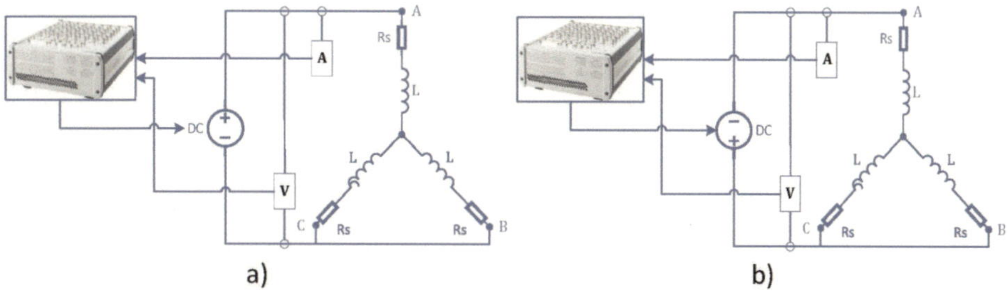

Figure 9.
d-axis inductance measurement circuit. (a) Positive supply phase A. (b) Negative supply phase A.

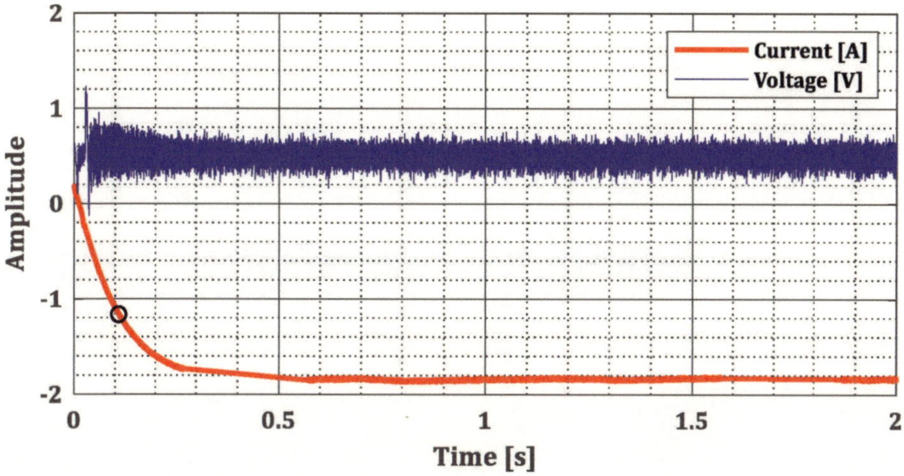

Figure 10.
Voltage and current through the coils for d-axis determination.

The q-axis inductance L_q was determined using the same method as the d-axis inductance L_d, with the key difference being that the rotor was aligned with the q-axis.

The first step is to connect phase B to the positive terminal and phase C to the negative terminal of the DC source, while leaving phase A floating, as shown in **Figure 11a**. Once the rotor is aligned with the q-axis, the shaft is mechanically locked, and the supply configuration is adjusted according to **Figure 11b**. Next, a voltage step is applied to the machine, and the current response is measured using a current probe. From the resulting current waveform, the time constant can be determined, allowing for the calculation of the q-axis inductance L_q.

The voltage and current through the coils are shown in **Figure 12**. A notable observation is that the current is higher for the same voltage step. This occurs because the rotor is unaligned with phase A, causing the coils to draw more current in order to generate a stronger magnetic flux and align the rotor with phase A.

Due to the increased current in the coils when measuring the q-axis inductance, the maximum step voltage was limited to 6.5 V, resulting in a current of 26.24A. In both cases, the current exceeds the nominal operating point of the machine by 20%.

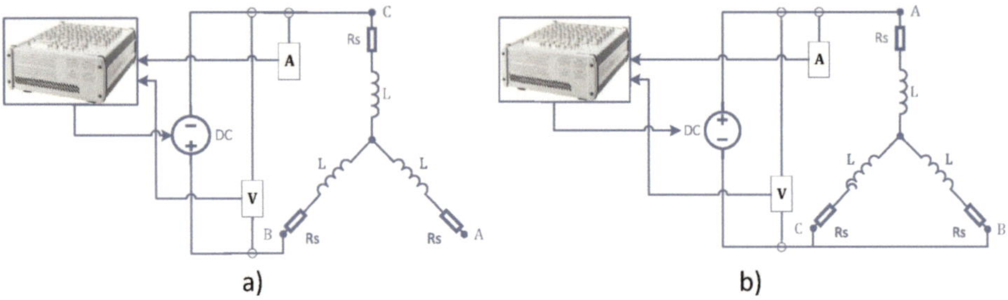

Figure 11.
q-axis inductance measurement circuit. (a) Negative supply phase C. (b) Positive supply phase C.

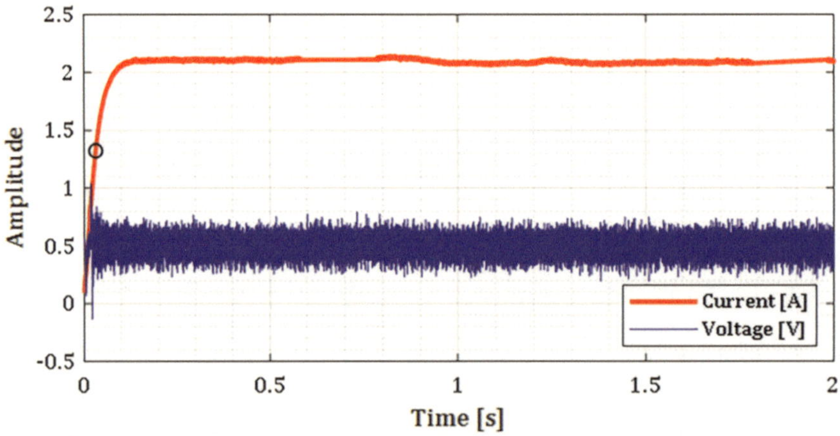

Figure 12.
Voltage and current through the coils for q-axis determination.

The d-q axis inductance measurements for all voltage steps are shown in **Figure 13**, compared with the simulation results of the prototype. A significant discrepancy can be observed between the measured and simulated d-axis inductance, particularly at low currents. The q-axis inductance slowly decreases in both cases by increasing the current.

4. Conclusions

This chapter began with an overview of low-power electric machines used in propulsion systems. It highlighted key aspects such as motor requirements, application constraints, and motor placement solutions. The concepts of modularity, modular, and segmented rotor designs were explored, along with their advantages and disadvantages. Among these, the modularity concept offers higher fault tolerance compared to modular and segmented rotor constructions. The electric machine is fault tolerant to both mechanical and electrical failures. In cases where a failure does not mechanically stop rotor movement or interrupt power supply, the electric motor will continue to work as long as at least one module is intact. The modular rotor design is easy to maintain and simple to disassemble. The current trend in electric machines is to avoid the use of rare earth materials. As such, the SynRM was chosen for this study as a viable alternative.

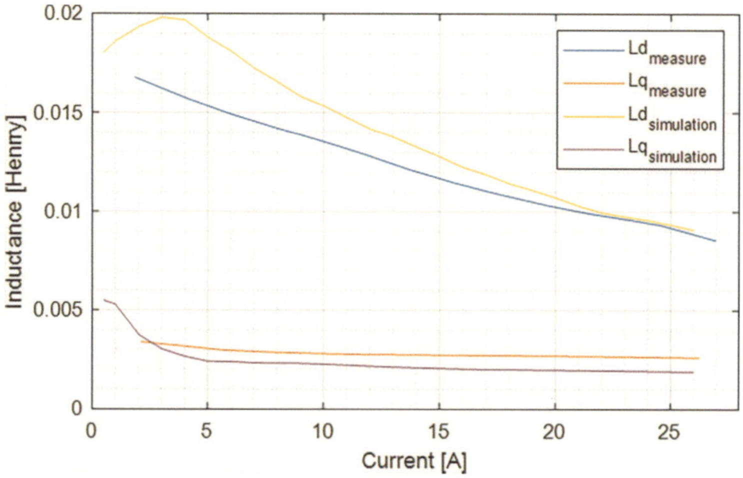

Figure 13.
d-q axis inductance tested on prototype and determined during the simulations.

The second section introduced the SynRM mathematical model and demonstrated that a higher saliency ratio and greater number of poles can enhance machine performance. The first study was on slots/pole combinations, starting with a high number of rotor poles to achieve greater electromagnetic torque. The initial assumption was that more poles would increase torque; however, results proved that a higher number of poles actually lead to lower developed torque. Thus, the 4-pole, 27-slot configuration was selected for further analysis.

The next study focused on increasing the saliency ratio by analyzing four different flux barrier shapes: round, rectangular, trapezoidal, and round-rectangular combination. Among these, the rectangular flux barrier proved to be the most balanced choice. Further analysis compared modular and modular-segmented rotor designs against classic structures. As a result, the modular rotor design was chosen for manufacturing.

In the final section of the chapter the results obtained from simulations were compared with experimental tests. The SynRM prototype was tested, and the main parameters (phase resistance, d-q axis inductances) were measured and analyzed. The chapter presented a methodology to determine SynRM parameters.

The difference between the simulations and actual measurements were higher at low currents for d-axis inductance, while remaining relatively consistent for q-axis inductance. This difference could be due to imperfections in the manufacturing process, specifically laser-cut steel sheets, which degrade material near the cutting margins and negatively impact electromagnetic performance.

The material presented in this chapter has the potential to serve as a foundation for future research, contributing to the development of an optimized ORSynRM that is suitable for use in different industrial and automotive applications.

Author details

Răzvan-Alexandru Inţe* and Florin-Nicolae Jurca
Technical University of Cluj-Napoca, Cluj-Napoca, Romania

*Address all correspondence to: razvan.inte@mae.utcluj.ro

References

[1] Wu D, Deng W. Characteristics analysis of urban traffic in small and medium-sized cities in context of mechanization–with Huai'an as an example. Procedia - Social and Behavioral Sciences. 2013;**96**:2240-2247. DOI: 10.1016/j.sbspro.2013.08.253

[2] Masson-Delmotte V, Zhai P, Pirani A, Connors SL, Péan C, Berger S, et al., editors. Summary for policymakers. In: Climate Change 2021: The Physical Science Basis. Contribution of Working Group I to the Sixth Assessment Report of the Intergovernmental Panel on Climate Change. Cambridge, United Kingdom and New York, NY, USA: Cambridge University Press; 2021. p. 2391. DOI: 10.1017/9781009157896

[3] European Commission. Commission Regulation (EU) No 692/2008 of 18 July 2008 Implementing and Amending Regulation (EC) No 715/2007 of the European Parliament and of the Council on Type-Approval of Motor Vehicles with Respect to Emissions from Light Passenger and Commercial Vehicles (Euro 5 and Euro 6) and on Access to Vehicle Repair and Maintenance Information (OJ L 171, 29.6). EU Monitor; 2008. Available from: https://www.eumonitor.eu/9353000/1/j4nvk6yhcbpeywk_j9vvik7m1c3gyxp/vi8rm2zwecvu

[4] Core Writing Team. Climate change 2014: Synthesis report. In: Pachauri RK, Meyer LA, editors. Contribution of Working Groups I, II and III to the Fifth Assessment Report of the Intergovernmental Panel on Climate Change. Intergovernmental Panel on Climate Change; 2015. 151 p. ISBN: 978-92-9169-143-2

[5] Hubka L. Electric cars in the Czech Republic – The analysis of CO2 emissions reduction. In: 2019 20th International Carpathian Control Conference (ICCC); Krakow-Wieliczka, Poland. 2019. pp. 1-6. DOI: 10.1109/CarpathianCC.2019.8765971

[6] ACEA. ACEA Proposal for Euro VII. Brussels, Belgium: European Automobile Manufacturers' Association; 2021. [Online]. Available from: https://www.acea.auto/files/ACEA_position_paper_Euro_VII_proposal.pdf

[7] ACEA. ACEA Proposal for Euro 7. Brussels, Belgium: European Automobile Manufacturers' Association; 2021. [Online]. Available from: https://www.acea.auto/files/ACEA_position_paper_Euro_VII_proposal.pdf

[8] Stilo L, Segura-Velandia D, Lugo H, Conway PP, West AA. Electric bicycles, next generation low carbon transport systems: A survey. Transportation Research Interdisciplinary Perspectives. 2021;**10**:100347. ISSN 2590-1982. DOI: 10.1016/j.trip.2021.100347

[9] Amelia B, Huang HW, de Haas M. Assessing the E-bike trends and impact on sustainable mobility: A national-level study in the Netherlands. Journal of Cycling and Micromobility Research. 2024;**2**:100027. DOI: 10.1016/j.jcmr.2024.100027, ISSN 2950-1059

[10] Hosseini K, Choudhari TP, Stefaniec A, O'Mahony M, Caulfield B. E-bike to the future: Scalability, emission-saving, and eco-efficiency assessment of shared electric mobility hubs. Transportation Research Part D: Transport and Environment. 2024;**133**:104275. DOI: 10.1016/j.trd.2024.104275, ISSN 1361-9209

[11] Kazemzadeh K, Bansal P. Electric bike level of service: A review and

research agenda. Sustainable Cities and Society. 2021;**75**:103413. DOI: 10.1016/j.scs.2021.103413, ISSN 2210-6707

[12] European Parliament and Council. Regulation (EU) no 168/2013 of the European Parliament and of the Council of 15 January 2013 on the approval and market surveillance of two- or three-wheel vehicles and quadricycles. Official Journal of the European Union. 2013;**L 60**:52-128

[13] Muetze A, Tan YC. Electric bicycles - a performance evaluation. IEEE Industry Applications Magazine. 2007;**13**(4):12-21. DOI: 10.1109/MIA.2007.4283505

[14] Heidari H, Rassõlkin A, Kallaste A, Vaimann T, Andriushchenko E, Belahcen A, et al. A review of synchronous reluctance motor-drive advancements. Sustainability. 2021;**13**: 729. DOI: 10.3390/su13020729

[15] Felix Stix. The Best eMTB Motor of 2021 - The 8 Hottest Motors in Test. 2021. Available from: https://ebike-mtb.com/en/the-best-emtb-motor-review/

[16] Bostanci E, Moallem M, Parsapour A, Fahimi B. Opportunities and challenges of switched reluctance motor drives for electric propulsion: A comparative study. IEEE Transactions on Transportation Electrification. 2017;**3** (1):58-75

[17] Zabihi N, Gouws R. A review on switched reluctance machines for electric vehicles. In: 2016 IEEE 25th International Symposium on Industrial Electronics (ISIE); Santa Clara, CA, USA. 2016. pp. 524-529. DOI: 10.1109/ISIE.2016.7744992

[18] Suryadevara R, Fernandes BG. Control techniques for torque ripple minimization in switched reluctance motor: An overview. In: 2013 8th IEEE International Conference on Industrial and Information Systems (ICIIS); Peradeniya, Sri Lanka. 2013. pp. 24-29

[19] Jurca N-F, Inţe RA, Popa D-C, Vărăticeanu B, Minciunescu P, Marţiş C. Electromagnetic and mechanical analysis of a modular outer rotor synchronous reluctance machine for light propulsion vehicles. IEEE Transactions on Transportation Electrification. 2021;**7** (4):2798-2811. DOI: 10.1109/TTE.2021.3067135

[20] Inte R-A, Jurca F-N. Torque ripple reduction of outer rotor synchronous reluctance machine with concentrated winding. In: 2017 19th International Conference on Electrical Drives and Power Electronics (EDPE); Dubrovnik, Croatia. 2017. pp. 248-252. DOI: 10.1109/EDPE.2017.8123240

[21] Vaseghi B, Takorabet N, Caron JP, Mobarakeh BN, Meibody- Tabar F, Humbert G. Study of different architectures of fault-tolerant actuator using a two-channel PM motor. IEEE Transactions on Industry Applications. 2011;**47**(1):47-54

[22] Babetto C, Bianchi N, Torreggiani A, Bianchini C, Davoli M, Bellini A. Optimal design and experimental validation of a synchronous reluctance machine for fault-tolerant applications. In: Proc. IEEE Energy Conversion Congress and Exposition (ECCE). 2019. pp. 4880-4887

[23] Baek J, Bonthu SSR, Choi S. Design of five-phase permanent magnet assisted synchronous reluctance motor for low output torque ripple applications. IET Electric Power Applications. 2016;**10**(5): 339-346

[24] Li GJ, Ma XY, Jewell GW, Zhu ZQ. Novel modular switched reluctance machines for performance

improvement. IEEE Transactions on Energy Conversion. 2018;**33**(3): 1255-1265

[25] Zhao W, Lipo TA, Kwon B-I. A novel dual-rotor, axial field, fault-tolerant flux-switching permanent magnet machine with high-torque performance. IEEE Transactions on Magnetics. 2015; **51**:1-4. DOI: 10.1109/tmag.2015.2445926

[26] Inţe RA. Outer rotor synchronous reluctance machine for light electric propulsion system [PhD thesis]. Department of Electrical Machines and Drives, Technical University of Cluj Napoca; 2023

[27] Hwang S-H, Kim J-M, Van Khang H, Ahn J-W. Parameter identification of a synchronous reluctance motor by using a synchronous PI current regulator at a standstill. Journal of Power Electronics. 2010;**10**. DOI: 10.6113/JPE.2010.10.5.491

[28] Bobek V. PMSM Electrical Parameteres Measurement. Document number AN4680, Rev. 0 02/2013

Chapter 2

Perspective Chapter: Fundaments on Fault Tolerance for Electrical Machines' Structures

Mircea Ruba, Claudia Marțiș, Paula Șerban and Sebastian Ursache

Abstract

The chapter surveys several types of electrical machines from the perspective of fault tolerance, namely induction, switched reluctance, and permanent magnet synchronous machines. It begins with a brief discussion of the induction machine, offering solutions that enhance continuous operation despite fault occurrences. Given that switched reluctance machines have inherent design and structural benefits for fault-tolerant operation, this type of machine receives extended attention. Various machine structures are discussed, ranging from classical designs to those with increased stator-to-rotor pole ratios, and further to modular configurations, multi-axis movement systems, and linear designs. For each, fault-free and fault-present operation scenarios are analyzed and explained. The chapter ends with a multiphase permanent magnet synchronous machine configured as a starter-alternator, which, based on results, is shown to be a good candidate for continuous operation in case of fault occurrence.

Keywords: fault-tolerant electrical machines, fault-tolerant control, fault compensation, induction machine (IM), switched reluctance machine (SRM), permanent magnet synchronous machine (PMSM)

1. Introduction

Even in fuel-powered automobiles, the number of electrical machines and drives onboard is significant, as they play a crucial role in both operational and comfort-related functions. This study focuses on electrical machines used in the electric vehicle (EV) industry, for traction, starter–alternator systems, and auxiliary device motors [1–4].

This chapter reviews common drive solutions for the automotive industry, considering the need for reliable operation under demanding conditions while maintaining high energy efficiency [5]. The following drive types are examined: the induction machine (IM), various switched reluctance machine (SRM) configurations, and the permanent magnet synchronous machine (PMSM), along with a study on rotary-linear moving machines, all of which are highly attractive from a fault-tolerant operation perspective [5–7]. While permanent magnet (PM) direct current (DC) motors are commonly used in traction applications (primarily in micro-motor applications), this study focuses on multiphase motors designed for the automotive industry [2].

The chapter briefly introduces potential fault sources in the machine-converter assembly and explores strategies to mitigate unbalanced operation while maintaining desired drive performance, such as smooth torque and speed, as well as improved efficiency through reduced iron and copper losses.

To evaluate the selected electrical drives, extensive numerical and analytical computations were performed using specialized software. Finite element method (FEM) analysis was conducted via Flux 2D–Simulink coupling [8], assessing operation under fault-free and faulty conditions in both motor and generator modes, using various control and converter strategies [9–12].

A fault-tolerant drive refers to a machine-converter system capable of maintaining satisfactory operation even after a fault occurs. Potential faults can be categorized into two main groups: electrical machine faults—such as winding malfunctions—and power electronics faults, including transistor failures and filtering capacitor breakdowns [3, 4, 13]. The goal is to obtain a fault-free system capable of continued functionality despite any of these malfunctions.

The chapter offers a brief literature review of fault-tolerant systems, particularly those applied in the automotive field, focusing on three main types: the induction machine (IM) [13], the switched reluctance machine (SRM) [14, 15], and the permanent magnet synchronous machine (PMSM) [16].

The IM remains the industry's preferred choice due to its simple construction and control. For squirrel cage induction motors (SC-IM), various winding and coil faults are analyzed.

The SRM drive, with its concentrated phase windings, is a favored topology among researchers for fault-tolerant operations. However, thorough fault characterization remains critical. This study examines open-circuit and short-circuit faults, along with the impact of different phase and pole configurations.

The PMSM is increasingly considered a viable replacement for IMs in traction applications. This study evaluates the influence of phase and pole numbers as well as various drive-related faults. Additionally, it discusses the issue of irreversible demagnetization of PMs. One potential solution involves compensating for PM losses using an auxiliary winding [10]. Another solution [17] involves a combination of structural design and alternative core materials to improve both fault tolerance and cost efficiency. Evaluating the feasibility of such solutions is especially important for machines used in sensitive industries such as medical and aerospace applications. Exploring diverse topologies under fault conditions helps to identify optimal solutions for specific applications [18]. In addition, fault signatures can support the development of fault-aware control systems that enable post-fault continuous operation [19].

2. Analysis of electrical machines from the perspective of fault tolerance

The following analysis draws on results obtained via multiple simulations, including FEM, analytical modeling, and validation using measured results.

2.1 Induction machine-based systems

Four topologies of the SC-IM were studied, all derived from a shared core design featuring 2 poles, 24 stator slots, and 20 rotor bars. Each had a rated torque of 80 Nm and 10 kW of power. Two winding configurations were analyzed: a double-layer three-phase and a six-phase layout, both implemented with either polar or shortened pitch designs (**Figure 1**). For each topology, a dedicated electrical circuit was modeled, including additional inductances to simulate end-winding effects.

FEM simulations were conducted under eight fault scenarios of increasing severity:

a. Healthy

b. Single coil short-circuit

c. Double coil short-circuit (three-phase topologies only)

d. Single-phase short-circuit

e. Single-phase open circuit

f. Single coil short-circuit (in a different phases)

g. Double coil short-circuit (across two phases)

h. Double-phase open circuit (six-phase topology only)

Partial coil faults were excluded, as full short-circuit scenarios already captured the maximum fault impact.

For a clearer comparison, the analysis here focus on the mean values of torque (**Figure 2**) and iron losses (**Figure 3**), excluding waveform ripples for presentation clarity.

Figure 1.
Types of winding solutions for the SC-IM: (I) distributed polar pitch; (II), (III) distributed shortened pitch; (IV) six-phase with distributed polar pitch.

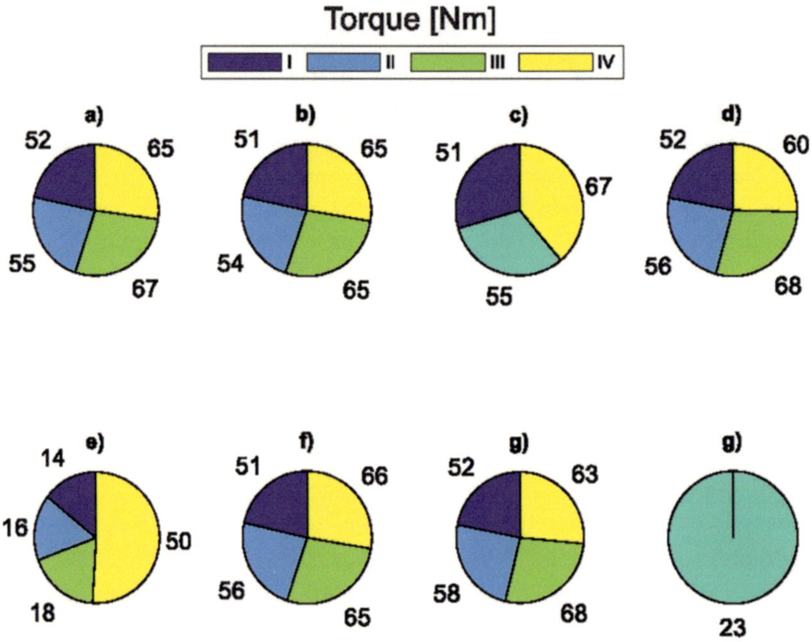

Figure 2.
The developed mean torque of the SC-IM.

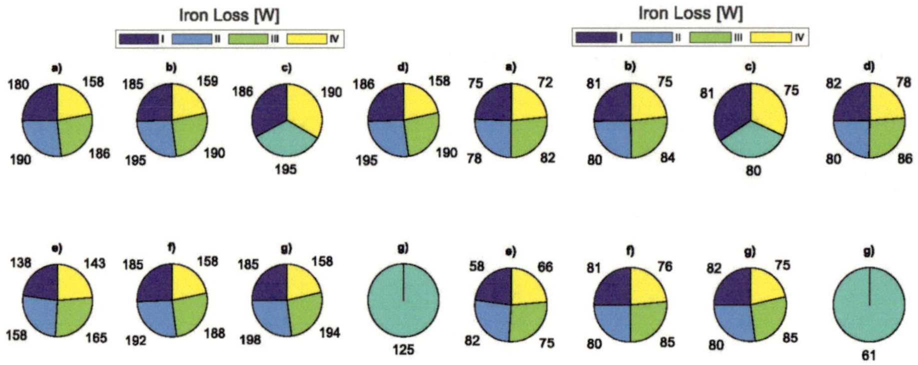

Figure 3.
Stator (left) and rotor (right) iron losses of the SC-IM.

For all cases, the geometry was identical, having a phase resistance of1.5 Ω, except the fourth topology (**Figure 1**), which used 0.75 Ω. The root mean square (RMS) voltage was set to 540 V (50 Hz) for the first three cases, while the fourth case was analyzed at 220 V RMS to achieve a mean torque value similar to that of the three-phase structure from the third case.

The computations were performed at a fixed speed (60% of the nominal value) to analyze a larger torque demand on the SC-IM. The results indicate that the three-phase shortened pitch winding (case III) provides better maximum torque performance than the polar winding (case I), with nearly the same iron loss levels due to reduced harmonic content. The six-phase topology demonstrated a more significant reduction in iron losses, partly due to the lower RMS voltage, which reduces power consumption and operational costs while requiring additional power sources and network equipment.

Under fault conditions (cases b–h), the mean torque and iron loss values remained relatively stable, though increased wave ripples were observed as fault severity increased, leading to higher energy losses. The most significant differences appeared in cases of open-circuit faults. The six-phase topology maintained a considerable torque capability even with a single-phase open circuit. In the case of a two-phase open circuit, the motor retained 40% of its mean torque from healthy operation.

To align an induction machine for fault-tolerant applications, specific design strategies must be considered. The stator design must have nonoverlapping concentrated windings, preferably in a single-layer shape. Redundancy can be introduced by repeating the three-phase system in the machine winding. In case of short circuit, the current can be limited by adding external inductances that slow down the slope of the current [20].

2.2 Switched reluctance machine-based systems

The switched reluctance machine (SRM) features a salient structure on both the stator and the rotor. The rotor is a passive electromagnetic element, typically built from laminated steel. The stator windings consist of coils wound around the stator poles, typically one coil per pole, with each machine phase comprising two coils connected in series or parallel. Since phases are energized individually, a controlled voltage source is required to obtain the desired excitation current. The SRM operates on the principle of variable reluctance [11]. Unfortunately, statistical data on SRM faults is not available in the literature. As a result, failure statistics from SC-IMs are often used as a reference [12]. In SC-IMs, the most common failures occur in the stator windings (30%) and bearings (40%) [13]. By extension, similar failure trends can be assumed for SRMs, suggesting that the most frequent failures in SRMs are likely winding faults or bearing failures.

Common types of winding defects include [13]:

i. Short circuit within a single coil of a phase

ii. Complete coil short circuit

iii. Complete short circuit of one phase

iv. One coil open circuit

v. Two-phase short circuit

vi. Ground (casing) short circuit

Winding faults typically result from factors such as mechanical vibration, excessive heat, aging, installation damage, or power converter failures [21–23].

To minimize the harmful effects of faults in systems utilizing SRMs, it is essential to detect these faults at the early stages. SRM winding faults can be detected using various failure detection methods. A basic approach involves using a simple fault detection device, such as the overcurrent detector, as shown in **Figure 4** (lefthand side). However, its effectiveness is limited due to a relatively slow response time and an inability to detect all fault types.

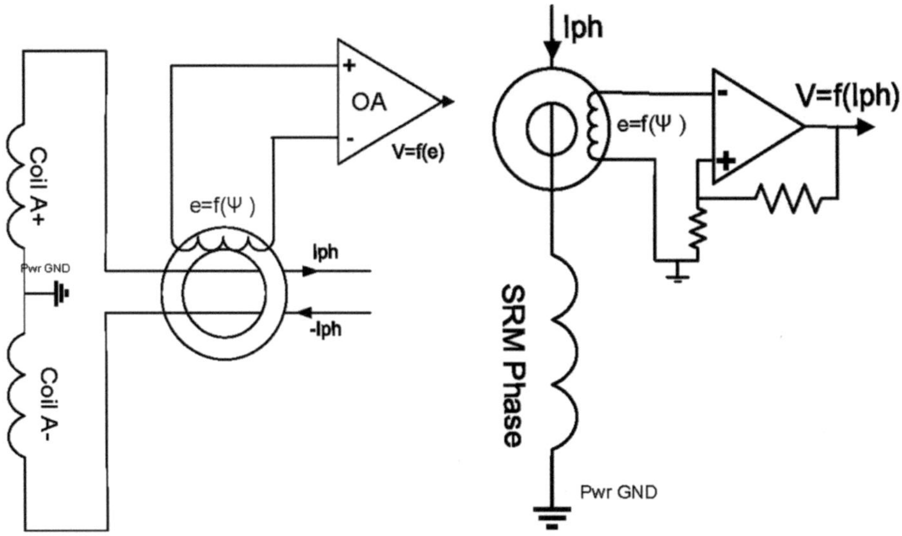

Figure 4.
Current difference detector (left) and rise-rate detector (right).

A more advanced fault detection method is the rise-rate sensing detector. It features a linear magnetic coupler encircling the upper phase-winding lead, with a core composed of a magnetic material that maintains linearity across the full operating range, as illustrated in **Figure 4** (righthand side). The coupler's output voltage corresponds to the time derivative of the phase current, which fluctuates due to winding faults. Any winding malfunctions in SRMs result in an asymmetrical flux distribution within the machine. The most effective way to highlight these changes is through a numerical field analysis of a sample SRM. The numerical field computations were performed using Flux 2D FEM software. The study was conducted for both a healthy machine and a faulty condition, where 20% of the turns in one phase winding were short-circuited. By conducting a time-stepping numerical field analysis, the magnetic flux variation in a single pole can be determined for different rotor positions. The results are then plotted as a function of the rotor's displacement, as shown in **Figure 5**. At position 0, the considered pole is completely unaligned with a rotor pole, and it aligns at around 30 degrees. The reduction in magnetic flux, caused by the shorted turns in coil A, is clearly evident. This decrease in magnetic flux can be considered as signature of the defect.

Dynamic simulations are among the most effective methods for highlighting the effects of different winding faults on SRM behavior. To this end, a conventional SRM was tested. Because accurate simulation requires both machine and converter modeling, two software environments were integrated: Flux 2D for finite element analysis, and MATLAB-Simulink for simulating the power converter and drive control.

The SRM model was constructed in Flux 2D (**Figure 6**). The FEM model of the SRM was integrated into the main Simulink program. The Simulink model of the drive system computes the power switch commands at each time step. At each time step, data is exchanged between Simulink and Flux 2D. The FEM model provides the main program with four essential signals: phase currents, electromagnetic torque, mechanical speed, and the SRM's angular position at a given time step. It is important to mention that mode-based analysis does not account for parametric variations due to manufacturing processes, operational conditions, temperature, voltage, and so on.

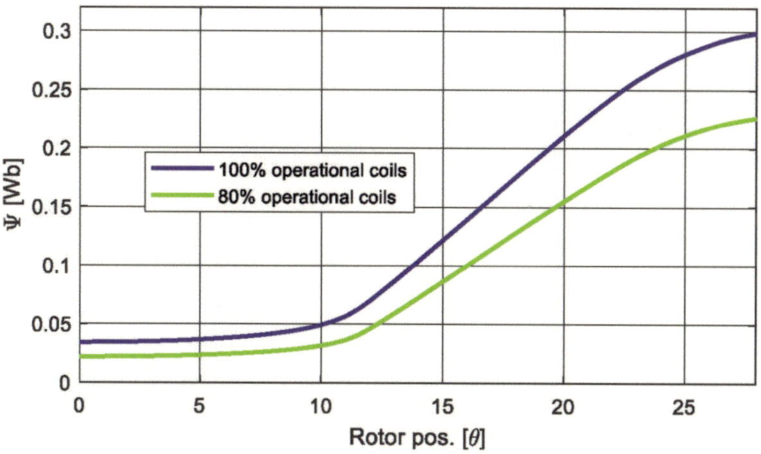

Figure 5.
Influence of short-circuit fault on magnetic flux.

Figure 6.
The classical SRM structure.

Five operating conditions were simulated:

a. Healthy machine (normal operation mode).

b. Two coils open from different phases.

c. Three coils open from different phases.

d. Four coils open from different phases.

As shown in **Figure 7(a)**, all the coils of the machine are operational and contributing to the torque development. In this case, the machine develops 12.1 Nm of mean torque. The ripples are typical for SRMs, and no smoothening method was applied

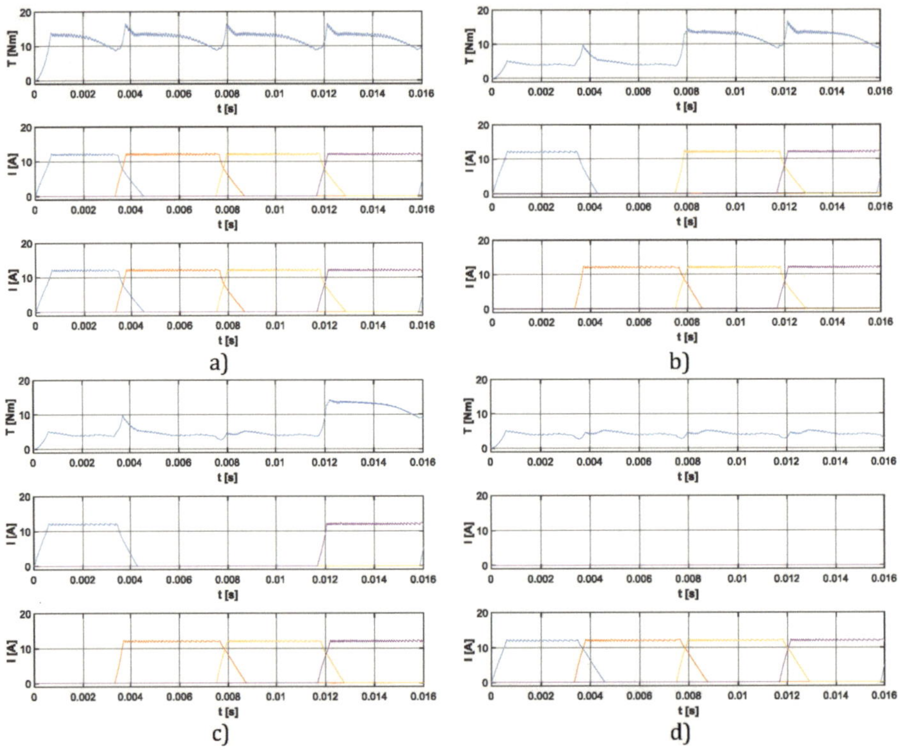

Figure 7.
SRM testing scenarios.

here because it falls outside the scope of this study. In **Figure 7(b)**, two coils from two different phases are faulted. In doing so, each affected phase contributes torque from only one active pole winding. Hence, only 83% of the rated torque is developed,; however, the machine remains operational.

In **Figure 7(c)**, the severity of the fault is increased, reaching three open-circuit coils from three different phases. Here, the machine's torque decreases to 48%of the rated value, reaching full torque only when the heathy phase is active.

The final tested condition involves four open-circuit coils (half of the machine's total winding count). In this case, the machine still develops 33% of its rated torque. However, because the remaining phases contribute equally, torque ripple is diminished.

In all the studied cases, one general observation is clear: parallel connection of phase coils already introduces a degree of fault tolerance in conventional SRM designs. Fault tolerance can be further improved through various methods, such as increasing the number of stator and rotor poles [14] or splitting the phase windings into independently powered coils, referred to as channels [14]. A modular SRM with enhanced fault tolerance is thoroughly examined in [16].

The highly fault-tolerant machine offers the fault-tolerant abilities of the classical SRM as well as flexibility when it comes to maintenance. For example, in the event of a coil failure, only the affected stator module, which includes both the iron core and winding, needs to be replaced **Figure 8** shows such a machine. The machine has five phases, each of which is divided into two channels. Each channel is wound

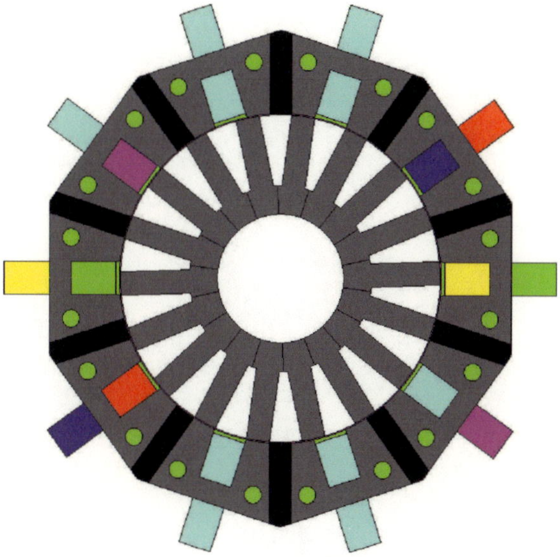

Figure 8.
Five-phase modular SRM.

on one of the eight modules' yokes. The modules of a single phase are placed diametrically opposite one another. The stator modules are separated by nonmagnetic spacers, which also ensure the correct shifting of the modules. Thanks to the reduced flux paths, the machine experiences lower losses compared to the traditional design. Moreover, the forces are more evenly distributed. The modular construction offers the benefits of easy manufacturing and quick replacement of damaged modules in the event of a winding failure. To replace a faulty module, it is only necessary to remove a single end shield and the two fixing rods of the damaged module, enabling easy extraction and replacement. This method eliminates the need to disconnect the machine from its load, offering a major advantage in industrial applications.

To analyze the fault tolerance of the proposed machine, simulations were performed under various operating conditions. Initially, a co-simulation approach was utilized, where the main program was implemented in Simulink, connected to the SRM from Flux 2D. The machine model was integrated into the main simulation via the Flux-to-Simulink interface.

Due to the lengthy computation times required, this approach was not feasible for simulating the machine within an advanced control system. Therefore, an alternative method was adopted for the dynamic simulation of the speed control system with the fault-tolerant SRM under investigation. This model relies on two key machine characteristics: torque and the magnetic flux through the energized coil, both of which are dependent on rotor position and current, as illustrated in **Figure 9**. These characteristics were initially derived through field computations using Flux 2D.

Figure 10 shows the main simulation schematic in Simulink. In this model, the data from **Figure 9** are included as lookup tables to replicate the accuracy of the FEM results even in a Simulink approach.

The same testing scenarios applied to the classical SRM were conducted on the modular version. **Figure 11** presents the corresponding results. To avoid redundancy, the key takeaway is that the machine maintains continuous operation across varying fault intensities. Although torque ripples is more pronounced compared to the

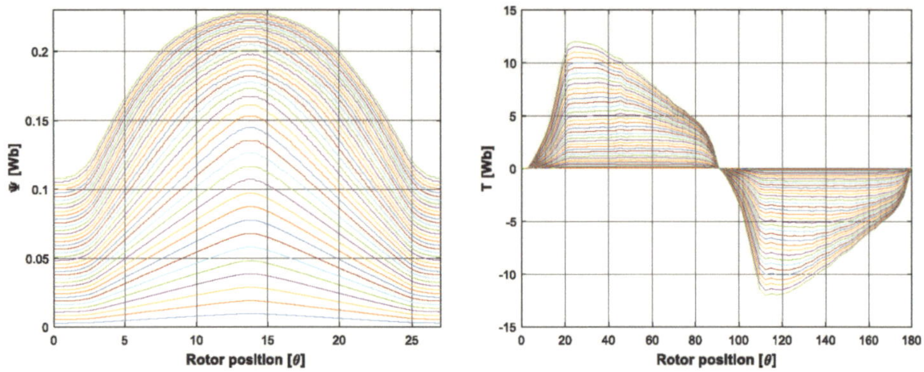

Figure 9.
The phase flux (left) and torque range (right) for different currents as a function of rotor position.

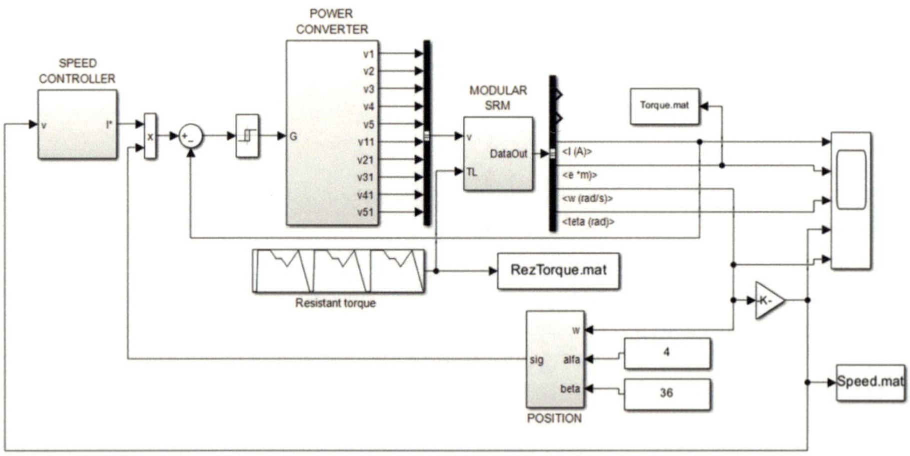

Figure 10.
The main MATLAB Simulink program for the modular SRM.

classical SRM, the modular design has the added value of simple assembly and disassembly when needed.

Creating a fault-tolerant version of a conventional electrical machine requires modifications to its topologies. To design an effective fault-tolerant electrical machine, it is essential to account for its losses as well. The primary design strategy focused on reducing the flux paths in the new motor structure, as shorter flux paths lead to lower iron losses. For the initial study, we selected a 12/8 SRM structure, as depicted in **Figure 12** (lefthand side).

The machine consists of 12 concentrated coils wound around each stator pole. To form a phase, two coils from diametrically placed poles are connected, resulting in a six-phase structure. This topology was chosen for its ease of adaptation into a fault-tolerant version. The proposed modifications involve adjustments to both the winding connections and the rotor structure.

The modified SRM features 14 rotor poles, as shown in **Figure 12** (righthand side). The key to achieving fault tolerance is to alter the flux path that crosses the air gap and closes between adjacent stator poles. In this rotor design, two adjacent pairs of stator poles contribute to torque development. Consequently, two adjacent

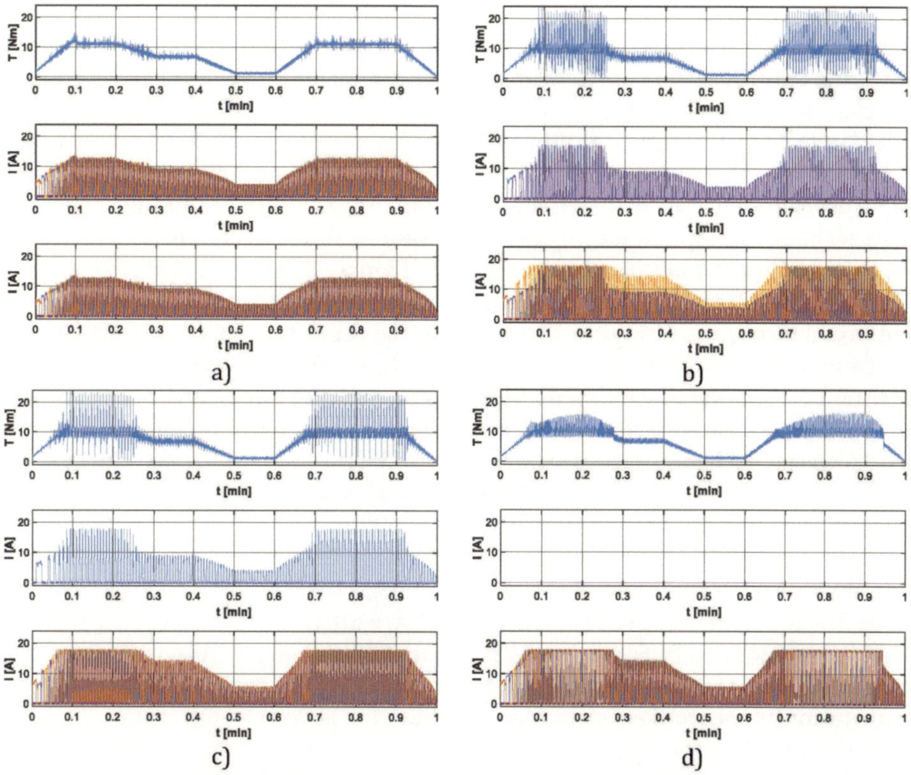

Figure 11.
Modular SRM fault-testing scenarios.

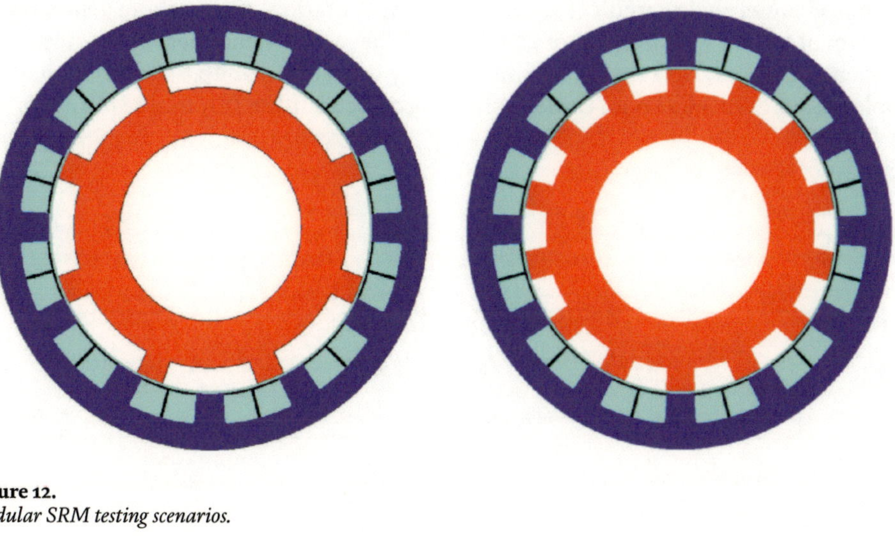

Figure 12.
Modular SRM testing scenarios.

phases energized simultaneously, activating a total of four coils at any given time. This winding arrangement enables the machine to continue operating even in the presence of faulty phases, while also minimizing torque ripple. The winding scheme follows a six-phase duplex configuration, where each phase is doubled. This means that phases are placed 180 mechanical degrees apart to ensure a balanced force distribution. In the event of a fault in one channel, the second channel continues to contribute to torque generation. By maintaining independent control and feeding for each channel, the system achieves inherent fault tolerance. Both machine variants under study share the same stator but differ in rotor design and phase/channel connections. Increasing the number of rotor poles enhances fault tolerance by improving torque development and ensuring safe operation under faulty conditions.

To evaluate this topology, four operating scenarios were analyzed: normal condition (**Figure 13(a)**), one channel open circuit (**Figure 13(b)**), two channels open circuit (**Figure 13(c)**), and one phase open circuit (**Figure 13(d)**).

A visual inspection of **Figure 13** underlines the benefits of this design. When one or even two channels of the winding are not working, the developed torque is still consistent with no significant increase in ripple across the torque profiles. In cases where a complete phase is lost, the torque that would be created by this phase drops to zero. However, such moments are usually passed by the system's inertia. Compared to its classical counterpart, the 12/14 SRM variant delivers 3–10% percent more torque during faulty scenarios with consistently lower ripples.

Another promising fault-tolerant SRM design is a 24/18 stator-to-rotor poles ratio. Here, the spatial arrangement of the stator and rotor teeth is optimized to maximize torque development while ensuring balanced force distribution. Each phase winding is divided into four coils. For example, in **Figure 14**, the coils corresponding to a single phase are wound around the stator poles that align with the rotor poles. This design ensures a continuous and efficient torque production process.

Due to the relatively high number of stator poles, torque ripple is significantly reduced, both in normal operation and under faulty conditions. This results in smoother performance, improved stability, and enhanced fault tolerance.

As shown in **Figure 15**, the 24/18 SRM demonstrates excellent performance across all tested fault conditions: normal operation (a), one channel open circuit (b), and one channel short-circuit (c). Under normal operation, the machine generates 46.7 Nm of torque with a ripple of 21.89 Nm. When a single channel experiences an open circuit (b), the torque remains virtually unchanged at 46.5 Nm, though ripple increases to 34.33 Nm. Under a short-circuit condition (c), torque output is

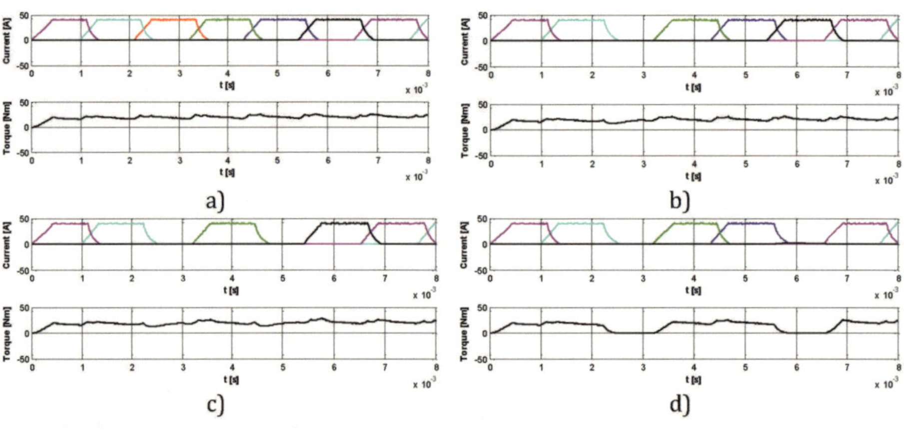

Figure 13.
Testing scenarios for the 12/14 SRM topology.

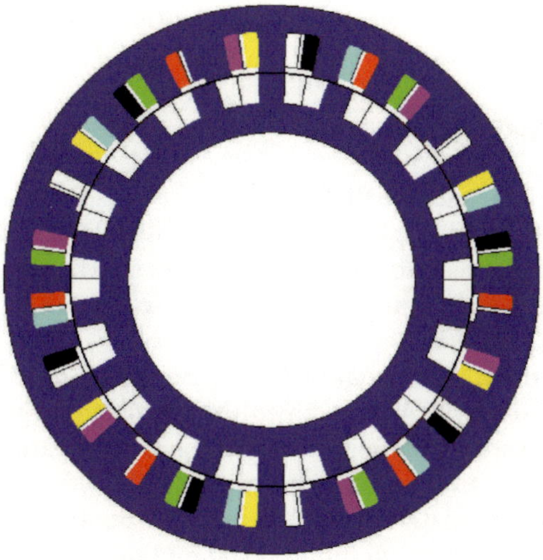

Figure 14.
Structure of the 24/18 SRM.

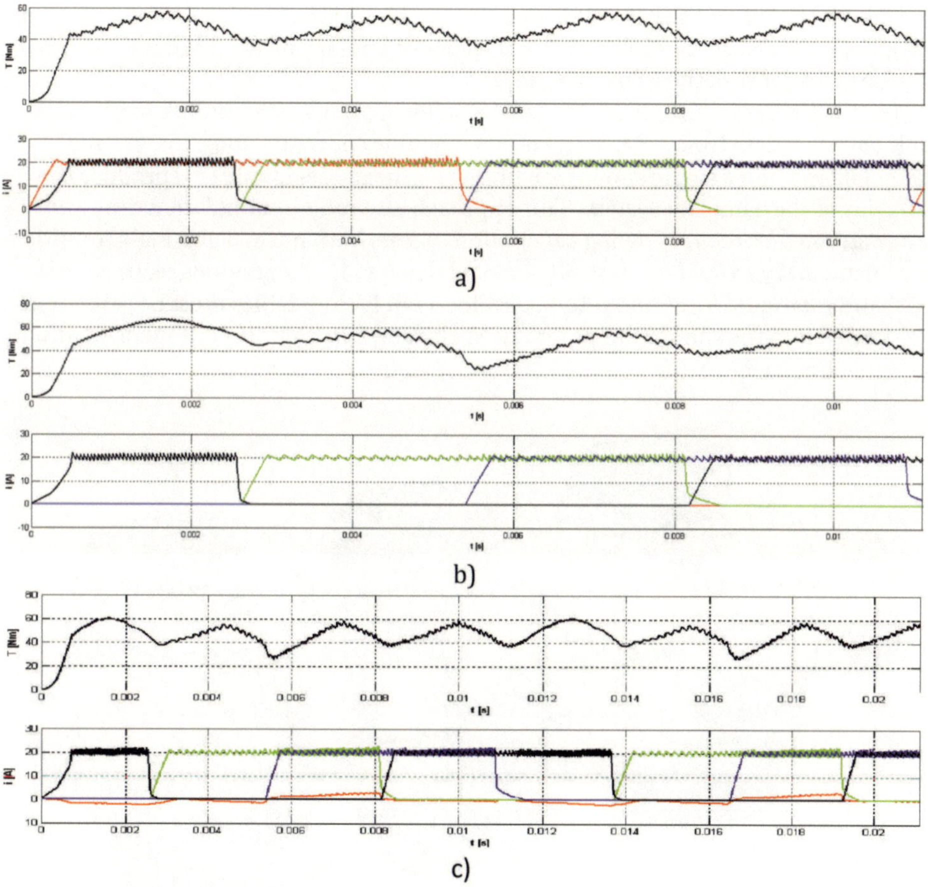

Figure 15.
Testing scenarios for the 24/18 SRM. (a) Normal condition (b) One channel open circuit (c) One channel short circuit.

maintained at 46.5 Nm, with ripple increasing slightly to 35 Nm. These results shows the machine's high fault tolerance.

2.3 Linear displacement-based machines

One custom electrical machine that by nature embeds the fault-free continuous operation is the limited axial displacement rotational SRM (LASRM). This machine combines the controlled axial movement of the main shaft using three stators and four rotors, all housed within a single machine. **Figure 16** presents a schematic representation of the LASRM.

The operational principle for axial displacement in the LASRM mirrors that of conventional rotational SRMs. As with any SRM, the machine generates torque by selectively energizing stator coils to guide the rotor along a path of minimum reluctance. In the LASRM, however, the current supply is shifted between multiple stators, causing the shaft to move axially between them while maintaining rotational motion. To preserve total torque output and induce axial motion, current distribution is balanced across the stators.

The LASRM is difficult to study in finite element analysis (FEA) due to its geometric complexity. Accurately modeling this machine requires 3D simulation and dual movement (radial and axial). While not impossible, it is extremely difficult to create a simulation scenario that can capture all types of operations for this design (both rotational and translational movement).

To effectively simulate the LASRM's behavior, a MATLAB Simulink model was built based on data from FEA for different machine configurations. Integrating these datasets into a single simulation program, the model is able to reproduce the behavior of the complete system. This approach also reduces simulation time and computational burden, allowing calculations across both normal and fault conditions. Complementary to the Simulink SRM model debated in the previous sections, here axial and rotational force and torque development lookup tables must be generated and loaded into the simulation program. Such data, depicted three-dimensionally, are shown in **Figure 17**.

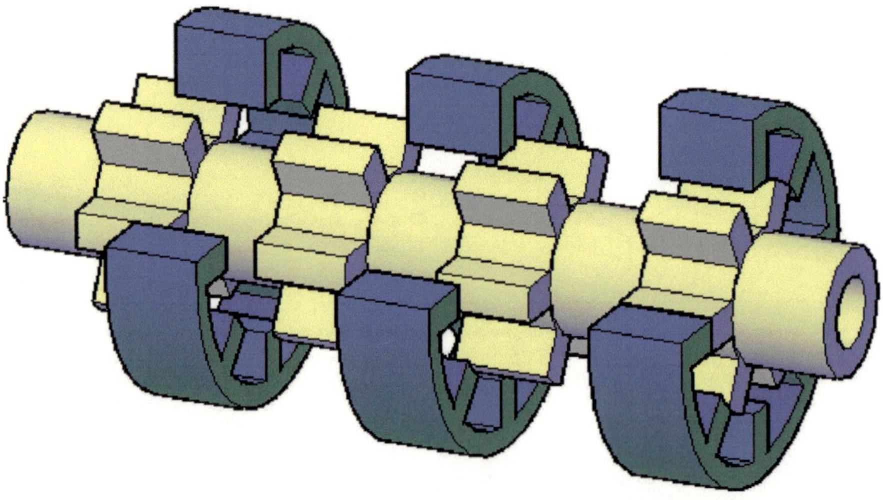

Figure 16.
LASRM architecture [24].

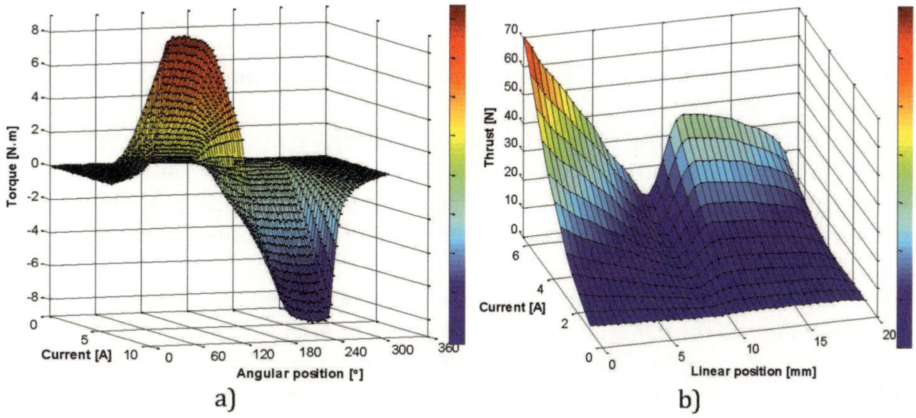

Figure 17.
Torque and force data from FEA models.

To analyze the dynamic behavior of the proposed LASRM under various conditions, the Simulink model was developed with a modular structure, as illustrated in **Figure 18**. The model clearly distinguishes the speed control block, motion controller, individual phase models of the power converters, stator stacks, and the mechanical system.

The primary simulation outputs include the phase currents in the windings, generated torque, axial movement, and the angular and linear displacements of the moving armature. Extensive studies were conducted using the developed simulation program.

In the following discussion, some of the most significant and insightful results are presented. To begin, the two types of motion—rotational and linear—were

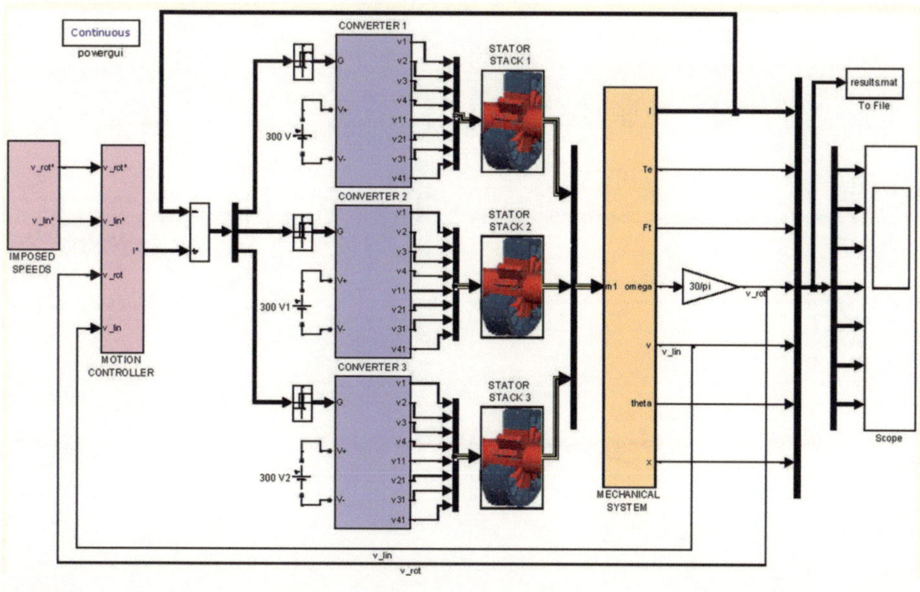

Figure 18.
MATLAB Simulink model of the simulator [24].

analyzed separately. The study began with rotational motion, focusing on a single stator and mover stack to enable an in-depth analysis. In the machine, only the poles of one stator stack can align with the mover poles along the axial direction at any given time.

Under normal operating conditions, all stator stacks of the LASRM contribute to total torque generation during rotation. Therefore, it it is also important to simulate the rotation with all three stator stacks active. **Figure 19(a)** presents the results obtained from rotational simulation, while **Figure 19(b)** presents the key results for the axial motion.

To avoid redundancy, detailed fault analyses are not reiterated. However, this structure combines the benefits of the multi-phase or multi-channel SRM detailed in the previous sections with fault-tolerant capabilities, offering also rotational and axial displacement. With appropriate control mechanisms, the system can maintain motion across both axes even in the event of a phase or coil failure. Remaining channels can temporarily handle increased current to compensate. The distributed, multi-coil structure enables operation even when only a subset of windings remains functional, providing adaptability across a broad range of fault scenarios.

2.4 Fault-tolerant modular linear motors

The structure of the studied linear modular motor is derived from that of the hybrid linear motor [25]. **Figure 20** shows a three-phase (six modules) version of this motor .

Each module comprises a permanent magnet, two salient-toothed poles, and a control coil. Every phase consists of two modules. **Figure 21** shows the operating principle of the modular linear motor. When the control coil is de-energized (**Figure 21(1)**), the permanent magnet's flux (Φpm) flows through the mid branch, as this path provides lower magnetic reluctance. In this condition, no significant force is generated.

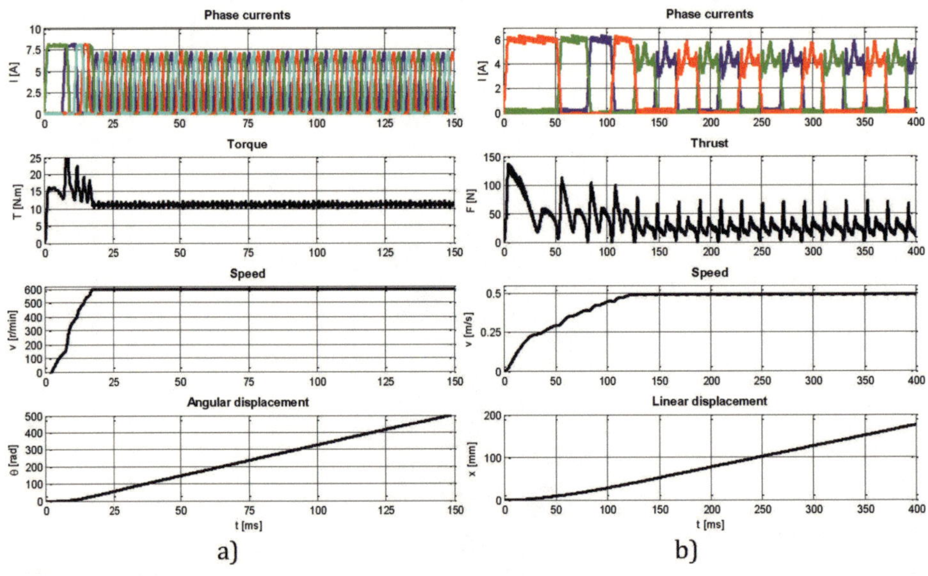

Figure 19.
Simulation results for rotational (a) and axial movement (b).

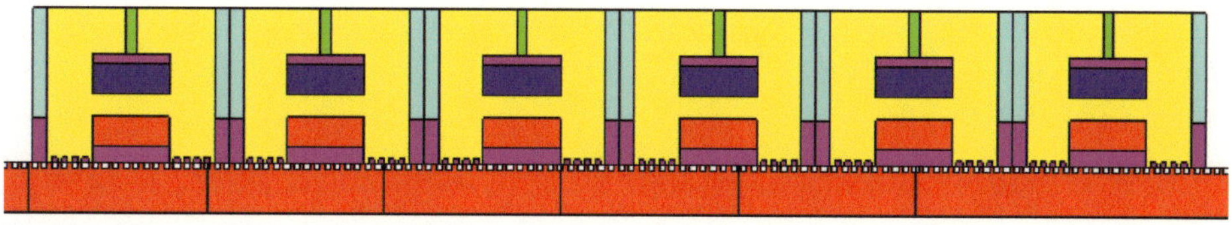

Figure 20.
Structure of the modular linear motor.

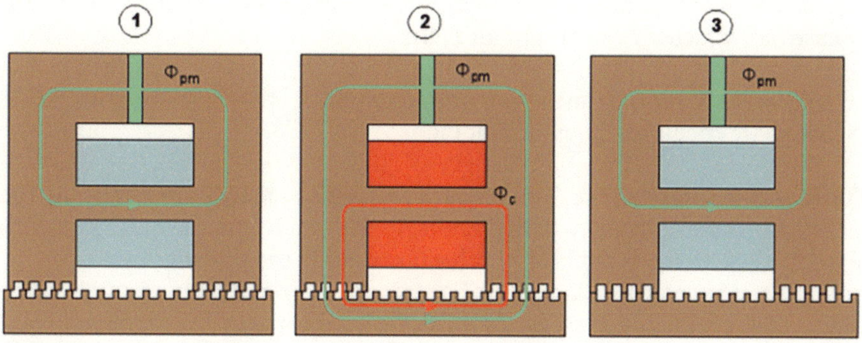

Figure 21.
Working principle of the modular linear motor [26].

When the control coil is energized (**Figure 21(2)**), the control flux generated (Φc) redirects the permanent magnet's flux through the air gap, resulting in the production of significant forces. Due to the tangential component of this force, the movable armature shifts by one step to minimize the air-gap magnetic energy, as shown in **Figure 21(3)**. The motor's resolution is determined by the tooth pitch and the number of modules. However, with advanced control strategies, the positioning resolution can be significantly enhanced.

The three-phase modular linear motor can be controlled using a variety of methods, ranging from simple open-loop control schemes to advanced, high-performance drive systems. In the latter case, the motor's speed can be adjusted from 0 up to several m/s. The speed is influenced by the motor design and is directly dependent on the current frequency. In addition, starting, breaking, and reversing the motor are straightforward to implement. When selecting the appropriate control strategy, it is important to balance the overall cost of the control system with the required performance capabilities. To maximize the average tangential force and minimize tangential force ripple, the control system must ensure that the excitation of the three modules' command coils is switched at a specific displacement—referred to as the optimal commutation angle (α_0)—before the system reaches intermediate equilibrium points. This angle can be computed depending only on the number of mover modules [27]. The modular linear motor is driven by an H-bridge power converter, with coil currents controlled using pulse-width modulation (PWM) techniques. Each module is equipped with its own H-bridge to supply the necessary current for generating the required force [28]. By connecting each H-bridge in parallel to the common main bus bars, any faulty windings can be completely isolated from the rest of the system. The converter's built-in intelligence facilitates this isolation process, allowing the faulty module to be disconnected as needed. Each H-bridge is individually controlled by managing the opening and closing of its power switches. This control is based on a strategy that utilizes the motor's linear position and the measured phase currents to optimize performance.

To evaluate fault resilience, four operational scenarios were tested:

a. Normal operating mode – all modules function without faults

b. Open circuit of one channel – a single channel experiences an open-circuit fault

c. Open circuit of two channels from different phases – two channels from separate phases are affected by open-circuit faults

d. Open circuit of two channels from the same phase – two channels within the same phase encounter open-circuit faults

The first case (**Figure 22(a)**) serves as the reference scenario, representing the motor operating without any winding faults. In this condition, the mean tangential force is 110.93 N, which is double the value observed in the previous case.

In the second scenario (**Figure 22(b)**), an open-circuit fault occurs in one channel. This results in a decrease in the tangential force to approximately 50 N, reflecting the loss of contribution from the faulty module. The mean tangential force in this case is 92.54 N, which corresponds to 84% of the force generated by the healthy motor.

The third case (**Figure 22(c)**) examines the effect of open-circuit faults in two channels from different phases. Here, the tangential force drops to around 50 N during two periods, leading to a mean tangential force of 73.50 N, which is roughly 66% of the force produced under normal operating conditions.

The final scenario (**Figure 22(d)**) investigates the most severe fault condition: open-circuit faults in two channels belonging to the same phase. In this case, the tangential force exhibits significant ripple, with a mean tangential force of approximately 74 N. Despite the severity of this fault, the motor still generates a considerable tangential force, comparable to the third case (**Figure 23**).

This study demonstrates that dividing each phase into two independent modules, effectively separating them into distinct channels, and adapting the winding configuration and control strategy yields a robust, fault-tolerant linear drive system.

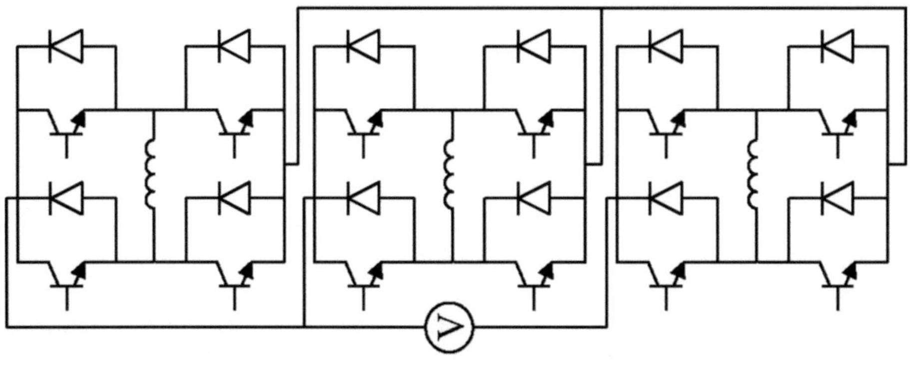

Figure 22.
The power electronics for the three-phase modular linear motor [26].

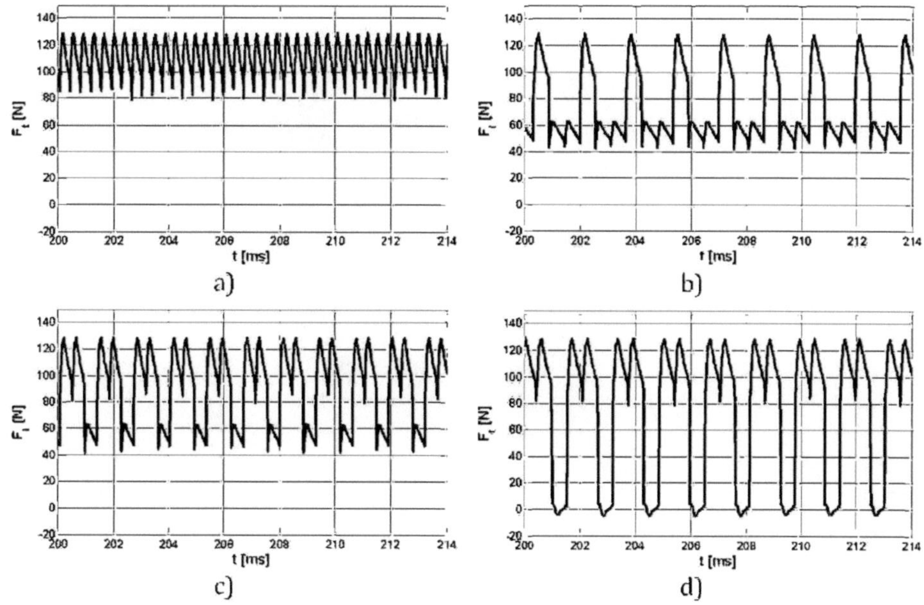

Figure 23.
Results under different operational conditions [26].

2.5 Fault-tolerant permanent magnet motors

The permanent magnet synchronous machine (PMSM) under study, depicted in **Figure 24**, was designed for starter-alternator applications. It features 9 phases and 16 poles, with each phase consisting of concentrated coils wound around each pole. The machine has a top power of 11 kW and operates at a rated voltage of 48 V DC in starter mode, with a peak current of 275 A. The machine's currents are regulated using pulse width modulation (PWM) techniques. Since the machine operates in both motor and generator modes, it has two rated speeds: 300 rpm in motor mode and 1000 rpm in generator mode.

The system generates the nine currents through a complex inverter with nine H-type bridges, each of which is independently controlled. To improve the inverter's fault tolerance, each bridge is connected in parallel with the main power bus bars. The machine was analyzed using FEM by developing a model in Flux 2D software. The system model encompasses both the converter and the rectifier for operation in both modes. For intelligent control of the machine, the Flux 2D-to-Simulink technology was employed. The command for each phase is calculated in a closed-loop system based on current regulation. To assess the machine's capability to function despite faults, various fault scenarios were analyzed in both motor and generator modes.

The study scenarios included:

a. Healthy mode

b. Single-phase fault

c. Double-phase fault

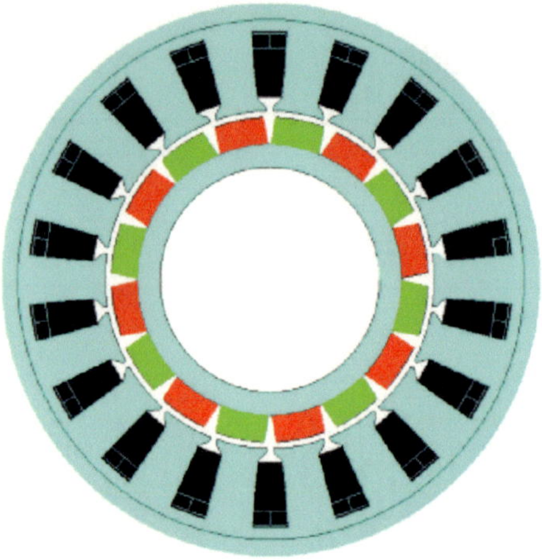

Figure 24.
The nine-phase PMSM.

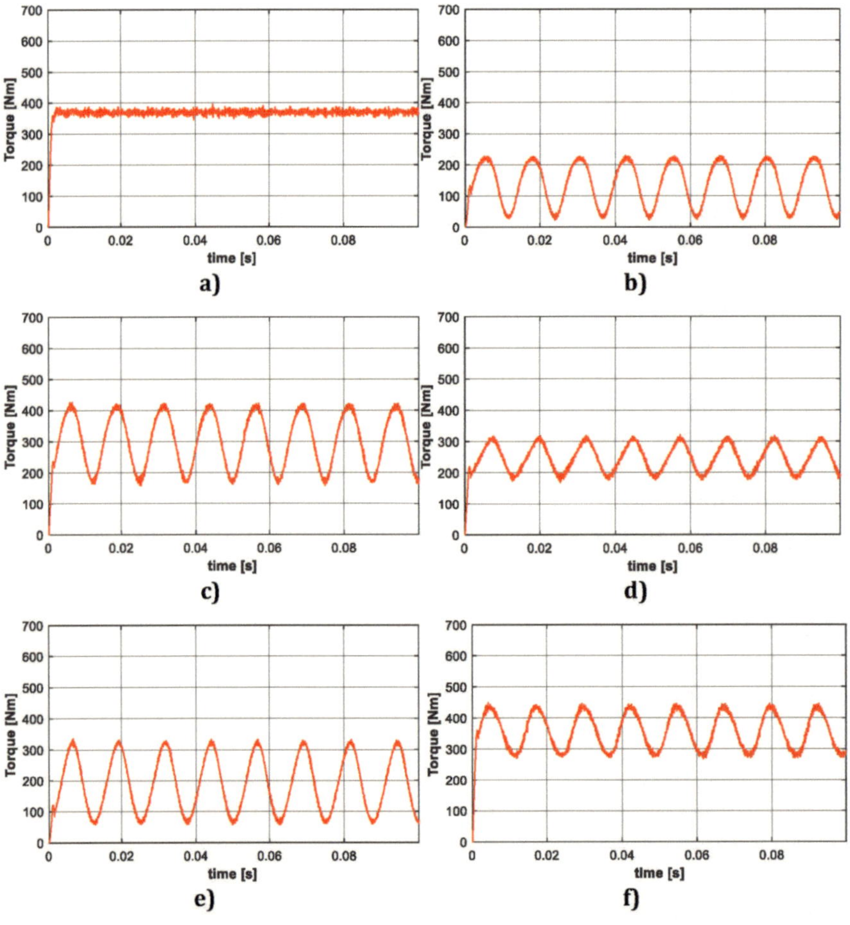

Figure 25.
Results of the nine-phase PMSM under different operational conditions.

Case	Torque [Nm]	Percent [%]	Ripple [Nm]
Normal	365	100	35
One faulted phase	355	96.8	105
Two faulted phases	300	81.3	210
Three faulted phases	245	66.5	90
Four faulted phases	190	51.4	265
Five faulted phases	135	34.2	185

Table 1.
Developed torque values for all studied cases.

d. Triple-phase fault

e. Four faulted phases

f. Five faulted phases

The motor regime was studied exclusively at 300 rpm for all cases. A starter's primary requirement is to generate sufficient torque to crank the combustion engine until it can run independently. All scenarios are referenced to healthy operation, with the corresponding values illustrated in **Figure 25**. In healthy (normal) operating conditions the machine can produce a torque of 365 Nm. As shown in **Figure 25**, the high number of phases results in a stable torque output with minimal ripple. However, as the severity of malfunctions increases—indicated by a higher number of faulty phases—the generated torque decreases, while torque ripple becomes significantly more pronounced.

Table 1 presents the mean values of the developed torque across all studied cases to highlight the machine's behavior under fault conditions. In the case of a single-phase open-circuit fault (**Figure 25(b)**), the machine can still produce torque close to its rated value. However, the absence of current contribution from the faulty phase results in increased torque ripple. In cases of two and four missing phases (**Figure 25(c)** and **(e)**), the torque exhibits the highest ripple values due to the insufficient number of phases to maintain a complete rotating magnetic field in the rotor. Conversely, with three and five open phases (**Figure 25(d)** and **(f)**), the torque ripple is reduced; however, this comes at the cost of a significantly lower developed torque.

3. Conclusions

This chapter traces the evolution of fault-tolerant strategies across several key electrical machine topologies. The three primary categories examined are the induction machine (IM), the switched reluctance machine (SRM), and the permanent magnet synchronous machine (PMSM).

While both induction and permanent magnet machines offer some degree of fault tolerance, their limited structural flexibility constrains their adaptability in highly fault-sensitive applications. In contrast, the SRM has significant potential for design modifications. When coupled with a proper power converter and control strategy, the SRM can withstand a wide range of fault conditions, from simple to complex, ensuring reliable continuous operation. The purpose of this chapter is to offer some general guidelines for selecting an appropriate fault-tolerant electrical machine, particularly for applications in the automotive industry.

Acknowledgements

This work was supported by a grant of the Ministry of Research, Innovation and Digitization, CCCDI UEFISCDI, project number PN-IV-P7-7.1-PED-2024-1479, within PNCDI IV.

Author details

Mircea Ruba*, Claudia Marţiş, Paula Şerban and Sebastian Ursache
Technical University of Cluj Napoca, Romania

*Address all correspondence to: mircea.ruba@emd.utcluj.ro

References

[1] Diallo D, Benbouzid ME-H, Makouf A. A fault-tolerant control architecture for induction motor drives in automotive applications. IEEE Transactions on Vehicular Technology. 2004;**53**(6):1874-1855

[2] Tirnovan R, Giurgea S, Miraoui A, Cirrincione M. Surrogate modelling of compressor characteristics for fuel-cell applications. Applied Energy (Elsevier). May 2008;**85**(5):394-403

[3] Welchko BA, Lipo TA, Jahns TM, Schulz SE. Fault tolerant three-phase AC motor drive topologies: A comparison of features, cost, and limitations. IEEE Transactions on Power Electronics. 2004;**19**(4):1108-1116

[4] Qin D, Luo X, Lipo TA. Reluctance motor control for fault-tolerant capability. In: Proceedings of the IEEE International Conference on Electric Machines and Drives, 18-21 May 1997. pp. WA1/1.1-WA1/1.6

[5] Atkinson GJ, Mecrow BC, Jack AG, Atkinson DJ, Sangha P, Benarous M. The analysis of losses in high-power fault-tolerant machines for aerospace applications. IEEE Transactions on Industry Applications. 2005;**42**(5):1162-1170

[6] Sawata T, Kjaer PC, Cossar C, Miller TJE, Hayashi Y. Fault-tolerant operation of single-phase SR generators. IEEE Transactions on Industry Applications. 1999;**35**(4):774-781

[7] Ede JD, Attalah K, Jiabin W, Howe D. Fault detection and fault-tolerant control of interior permanent magnet motor drive system for electric vehicle. IEEE Transactions on Magnetics. 2002;**38**(5):3921-3923

[8] Fodorean D, Djerdir A, Viorel IA, Miraoui A. A double excited synchronous machine for direct drive application - Design and prototype tests. IEEE Transactions on Energy Conversion. 2007;**22**(3):656-665

[9] Li S, Xu L. Strategies of fault-tolerant operation for three-level PWM inverters. IEEE Transactions on Power Electronics. 2006;**21**(4):933-940

[10] Fodorean D, Viorel IA, Djerdir A, Miraoui A. Performances for a synchronous machine with optimized efficiency while wide speed domain is attempted. IET Electric Power Applications. 2008;**2**(1):64-70

[11] Henneberger G, Viorel IA. Variable Reluctance Electrical Machines. Aachen (Germany): Shaker Verlag; 2001

[12] Nandi S, Toliyat HA. Condition monitoring and fault diagnosis of electrical machines - A review. In: Proceedings of the IEEE International Conference on Electric Machines and Drives (IEMDC'99), Seattle (USA). 1999. pp. 197-204

[13] Schinnerl B, Gerling D. Analysis of winding failure of switched reluctance motors. In: Proceedings of the IEEE International Electric Machines and Drives Conference (IEMDC'09), Miami (USA). pp. 738-743

[14] Jack AG et al. A comparative study of permanent magnet and switched reluctance motors for high-performance fault-tolerant applications. IEEE Transactions on Industry Applications. 1996;**32**:889-895

[15] Ruba M, Anders M. Fault tolerant switched reluctance machine study.

In: International Conference on Power Electronics, Intelligent Motion and Power Quality (PCIM'2008), Nürnberg (Germany). 2008

[16] Ruba M et al. Modular fault tolerant switched reluctance machine-design and dynamic simulations. In: 12th International Conference on Optimization of Electrical and Electronic Equipment (OPTIM'2010), Moieciu (Romania). pp. 441-446

[17] Sizonenko V, Vitek O. Preliminary design and comparison of 5 phase and 6 phase fault tolerant outer rotor permanent magnet synchronous machines with different electrical steel. In: 2022 PEMC, Brasov, Romania. 2022. pp. 390-395. DOI: 10.1109/PEMC51159.2022.9962852

[18] Wang B, Liu Y, Vakil G, Yang T, Zhang Z. Feasibility of permanent magnet fault tolerant machines for aircraft starter/generator systems. In: 2020 ICEM, Gothenburg, Sweden. 2020. pp. 2104-2110. DOI: 10.1109/ICEM49940.2020.9270965

[19] Tai Z, Zhang W, Jin H, Zhang J. Fault-tolerant control of dual three-phase permanent magnet synchronous machine under open phase fault. In: 2023 ICEMS, Zhuhai, China. 2023. pp. 1555-1560. DOI: 10.1109/ICEMS59686.2023.10344621

[20] Lee W, Choi G. A comprehensive review of fault-tolerant AC machine drive topologies: Inverter, control, and electric machine. In: 2021 SDEMPED, Dallas, TX, USA. 2021. pp. 269-275. DOI: 10.1109/SDEMPED51010.2021.9605560

[21] Hu Y, Tang Z, Blaabjerg F. Power losses analysis of converter for switched reluctance motor drive with fault-tolerant control. In: 2023 ICEMS, Zhuhai, China. 2023. pp. 3157-3162. DOI: 10.1109/ICEMS59686.2023.10344480

[22] Pires VF, Foito D, Pires AJ, Cordeiro A, Martins JF. A fault tolerant converter for a SRM drive based on a nine-switch inverter with a dual output. In: 2023 CPE-POWERENG, Tallinn, Estonia. 2023. pp. 1-6. DOI: 10.1109/CPE-POWERENG58103.2023.10227414

[23] Ali N, Gao Q, Ma K. Comprehensive cost evaluation of advanced fault-tolerant converters for switched reluctance motors. In: 24th International Conference on Electrical Machines and Systems (ICEMS), Gyeongju, Republic of Korea. 2021. pp. 344-348. DOI: 10.23919/ICEMS52562.2021.9634311

[24] Szabó L, Benţia I, Ruba M. Dual motion switched reluctance motor for advanced industrial applications. In: 2012 13th International Conference on Optimization of Electrical and Electronic Equipment (OPTIM), Brasov, Romania. 2012. pp. 544-549. DOI: 10.1109/OPTIM.2012.6231831

[25] Viorel IA, Szabó L. Hybrid Linear Step-per Motors. Cluj-Napoca: Mediamira Publisher; 1998

[26] Ruba M, Szabo L, Hopper E. FEM based studies on fault tolerant modular linear motors. In: 2009 PCIM, Nuremberg, Germany

[27] Szabó L, Ruba M, Fodorean D. Study on a simplified converter topology for fault tolerant motor drives. In: Proceedings of the 11th International Conference on Optimization of Electrical and Electronic Equipment (OPTIM'2008), Braşov (Romania). 2008. pp. 197-202

[28] Pires VF, Pires AJ, Foito D, Chaves M, Cordeiro A. A fault tolerant multilevel converter topology for an 8/6 SRM drive based on a cross-switched configuration. In: 2022 ESS, Kyiv, Ukraine. 2022. pp. 335-340. DOI: 10.1109/ESS57819.2022.9969238

Chapter 3

On Fault Tolerance of Data Storage Systems: A Holistic Perspective

Mai Zheng, Duo Zhang and Ahmed Dajani

Abstract

Data storage systems serve as the foundation of modern digital society. The enormous amount of data generated by people continuously makes fault tolerance for storage systems increasingly important. Unfortunately, modern storage systems consist of complicated hardware and software layers interacting with each other, which may contain latent bugs that elude extensive testing and hurt the data integrity once triggered. In this chapter, we take a holistic view to introduce the typical architecture and major components of modern data storage systems (e.g., solid-state drives, persistent memories, local file systems, and parallel/distributed storage systems at scale). Next, we discuss the representative bug detection and fault tolerance techniques across layers with a focus on issues that affect system availability and data integrity. Finally, we conclude with open challenges and future research directions.

Keywords: data storage systems, file systems, system failure, data loss, metadata corruption, crash consistency, fault injection, fault tolerance, reliability, resilience, security

1. Introduction

Data storage systems (DSS) play an essential role in modern society. The enormous data generated by various use cases (e.g., financial transactions, medical records, and scientific datasets) continuously makes the fault tolerance of data storage systems increasingly important.

Unfortunately, building a fault-tolerant storage system is challenging due to the ever-growing complexity. Modern storage systems consist of complicated hardware and software layers interacting with each other, which may contain latent bugs that elude extensive testing and hurt the data integrity once triggered. In particular, latent defects in fault-tolerance mechanisms can lead to severe consequences, including server downtime, data corruption, and financial losses. As data storage systems continue to evolve and grow in terms of scale and complexity, the risk of failures becomes increasingly prevalent [1–3]. In other words, understanding the system architecture and analyzing the fault tolerance thoroughly is imperative.

In this section, we introduce the typical architecture of modern data storage systems in order to lay the foundation for further discussions on storage fault tolerance (Section 2). As shown in **Figure 1**, a typical data storage system (DSS) mainly consists

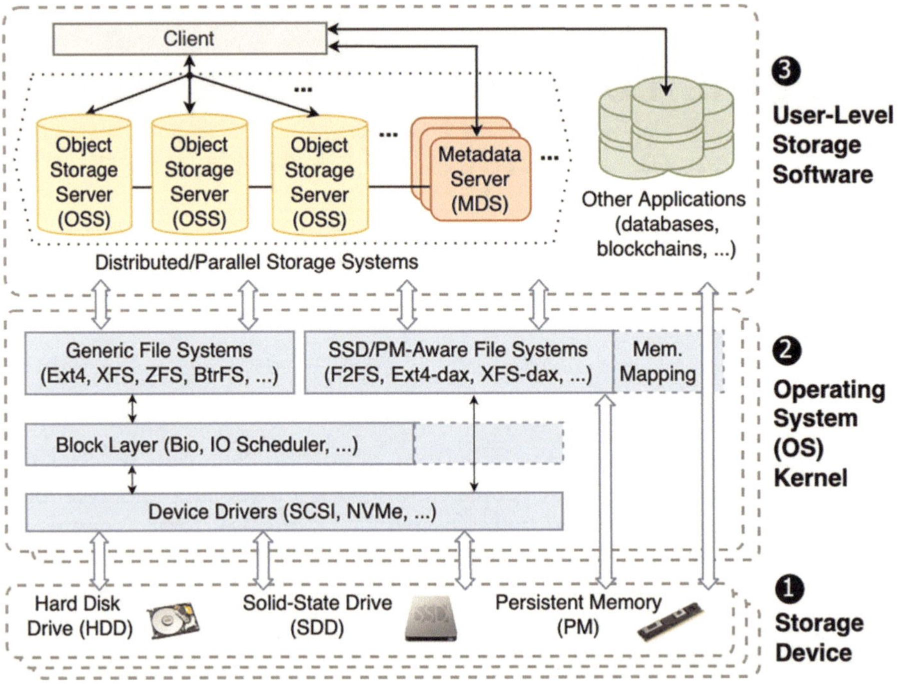

Figure 1.
A holistic view of data storage systems.

of three layers: storage device, operating system (OS) kernel, and user-level storage software. Each of them plays a unique role in managing data in computers, and the combination of them provides end-to-end data support, from handling user requests to interacting with hardware for various application scenarios. We elaborate on the three primary layers in the following sections.

1.1 Storage device: Persistent home for data

Storage device provides the necessary persistence to store data durably and serve as the foundation of data storage systems in practice. There are three main categories of storage devices today:

1.1.1 Hard disk drives (HDD)

HDDs have been the dominant storage medium for decades due to their cost-effectiveness and high capacity. HDDs use magnetic platters to store data, with a mechanical arm positioning a read/write head over spinning disks to access information. Despite their affordability and long lifespan, HDDs suffer from relatively slow access speeds due to mechanical latency and seek time. Typical rotational speeds range from 5400 to 15,000 RPM (revolutions per minute) today, leading to access times in milliseconds. HDDs are particularly well-suited for archival and bulk storage applications where cost per gigabyte is a priority. However, they struggle to meet the performance demands of modern workloads that require fast random access. To mitigate their limitations, techniques like caching and shingled magnetic recording (SMR) have been introduced to improve HDDs in terms of performance and/or capacity.

1.1.2 Flash-based solid-state drives (SSD)

SSDs have revolutionized storage by eliminating mechanical components, instead relying on NAND flash memory to store data electronically. Compared to HDDs, SSDs offer much lower latencies, with typical access times in the microsecond range. The absence of moving parts enables SSDs to deliver high-speed random reads and writes, making them ideal for performance-sensitive applications. SSDs come in various form factors, with NVMe-based drives offering significantly higher bandwidth than other traditional counterparts. However, NAND flash memory has inherent limitations, including limited write endurance and higher cost per byte compared to HDDs. To mitigate wear, SSDs employ techniques like wear leveling and garbage collection, which may introduce write amplification overhead. The transition from planar NAND to 3D NAND has improved density and endurance, helping SSDs achieve wider adoption across consumer and enterprise markets.

1.1.3 Persistent memory (PM)

PM technologies offer attractive features for developing storage systems and applications. Unlike traditional volatile memory technology (i.e., DRAM), PM provides non-volatility, allowing data to persist across power cycles while maintaining low-latency memory access. For example, Intel® Optane™ [4] can support byte-granularity accesses with latencies less than 3x of DRAM latencies [5], while also providing durability guarantees. These properties enable PM to serve as both a high-speed storage tier and an extension of main memory, bridging the performance gap between DRAM and SSDs. However, the PM also introduces new challenges, such as wear endurance limitations and the need for efficient software interfaces to leverage its byte-addressable capabilities. As persistent memory technologies continue to evolve, they hold promise for reshaping storage hierarchies and enabling new classes of high-performance applications.

Note that it is possible to organize multiple storage devices in an array to improve parallelism and redundancy, so as to achieve higher performance and/or fault tolerance (e.g., redundant arrays of inexpensive disks (RAID) [6]). In general, these storage devices including RAID-like solutions exhibit different characteristics and failure modes, which demand different strategies for ensuring fault tolerance in storage systems.

1.2 Storage software stack in OS: Managing data on a single computer

The storage software stack in the operating system (OS) kernel serves as an intermediary between user applications and physical devices to manage data on a single computer. We briefly introduce a few key components in the widely used Linux OS below, and refer interested readers to [7, 8] for more details.

1.2.1 File systems (FS)

File systems implement a set of file-related system calls (e.g., `open`, `read`, `write`, and `close`) to provide the file and directory abstraction to user applications. Internally, it formats the storage device into a set of data blocks and metadata structures (e.g., bitmaps and inodes) to ensure efficient management and access of data on device. Traditional file systems (e.g., Ext4 [9] and BtrFS [10]) have been designed for managing data on HDDs. More recently, new file systems have been proposed for new devices. For example, F2FS [11] is a special file system carefully designed for SSDs. Similarly, for persistent memory (PM), specialized file systems like NOVA [12] and PMFS [13] optimize performance by leveraging PM's byte-addressability. Moreover, traditional file systems including EXT4 and XFS have been extended with DAX (Direct Access) support, leading to EXT4-DAX and XFS-DAX. The DAX mode eliminates the page cache and bypasses the block layer, allowing applications to directly access persistent memory at near-DRAM speeds. In general, different file systems mainly differ in terms of their on-drive data structures and access methods, while their interfaces are largely the same following the POSIX standard [14]. Besides the basic file management, file systems may implement additional features such as journaling, copy-on-write (CoW), deduplication, snapshots, encryption, and compression, etc. to enhance reliability and security.

1.2.2 The block I/O layer

The block layer abstracts the physical storage medium, presenting a uniform block-based interface to upper-layer software. It manages data placement, scheduling, and I/O optimizations such as request merging and reordering. Linux's block I/O subsystem includes components like BIO, I/O schedulers, and the multi-queue block IO queuing mechanism (blk-mq) for high-performance devices like NVMe SSDs. Note that the Block I/O layer can be bypassed *via* DAX (Direct Access) mode, allowing applications to directly access memory-mapped storage without traditional block I/O overhead. In addition, the block layer can be integrated with storage virtualization techniques (e.g., Logical Volume Manager (LVM) and software RAID) to further enhance flexibility and resilience.

1.2.3 Device drivers

The device drivers directly communicate with storage devices based on standardized interfaces (e.g., SCSI, SATA, and NVMe). Each interface defines a set of commands and protocols for communications. In particular, modern NVMe drivers leverage multi-queue capabilities to maximize parallelism and reduce latency, significantly improving SSD performance. Persistent memory devices rely on drivers like Linux's `libnvdimm` to manage NVDIMM devices and expose them as either block devices or memory-mapped regions. Efficient driver implementation is crucial for ensuring low-latency, high-throughput storage access, particularly for emerging technologies like CXL-attached memory and computational storage.

1.3 User-level storage software: Enabling various applications at scale

The user-level storage software enables various application scenarios (e.g., distributed file systems, databases, and blockchains). In particular, parallel/distributed storage systems (DSS) are designed to manage data at scale [15–23]. They typically consist of a set of nodes with different functionalities, including *Management Server* to manage and store cluster-level configuration information, *Metadata Server* to manage and store the metadata, *Object Storage Server* to manage and store the actual user data as objects, and handle I/O requests, etc. Clients may interact with multiple storage servers responsible for managing user data, while metadata servers handle essential metadata operations such as file indexing and directory management.

Despite the similar functionalities, different DSSs may use different terminologies for their servers. For example, Luster is a parallel file system (PFS) widely used in high-performance computing (HPC) centers [19, 24], and it typically includes MDS (Metadata Server Node) and MDT (Metadata Target Node) for metadata and OSS (Object Storage Server) and (OST) Object Storage Target for user data. Similarly, Ceph is a distributed storage system designed to be highly scalable and fault-tolerant [20], which typically consists of one monitor/manager node (MON/MGR) and many OSD (Object Storage Daemon) server nodes.

Besides the basic functionalities, which are largely similar, different DSSs may introduce system-specific techniques to provide unique features. For example, the Ceph storage system organizes objects using a logical concept named *pool*. For better object management, objects in a Ceph pool are further divided into another logical group named placement group (PG). PGs reside on one or more OSD devices and can be overlapped on OSDs. On top of Ceph's object store service (RADOS), Ceph also integrates object gateway (RGW), block device service (RBD), and file storage service (CephFS). In addition, to ensure fault tolerance, DSS typically includes dedicated checking, recovery, and failure mitigation components. For example, Luster includes a fault-tolerance component called *LFSCK* to check and fix potential corruptions [25]. Similarly, Ceph supports erasure coding (EC) to provide redundancy of user data with low storage overhead. As of this writing, Ceph supports multiple EC plugins including Reed-Solomon (RS) codes and Clay codes *via* third-party libraries [26, 27].

Additionally, other storage-related applications (e.g., databases and blockchains) may exist at this level, offering structured data management and high-level storage abstractions for various application needs. Note that while the majority of the code is typically implemented at the user level, the storage applications may also include partial functionality directly in the OS kernel to improve performance. For example, Luster's *ldiskfs* backend is a variant of the Linux Ext4 file system that modifies Ext4 and relies on its extended attributes for metadata.

2. Fault tolerance analysis for data storage systems

2.1 Challenges

Given the complexity of data storage systems, as discussed in Section 1, achieving fault tolerance for valuable data is non-trivial in practice. In particular, storage systems do not live in isolation; there are implicit dependencies across layers in the architecture and ecosystem (**Figure 1**), which makes end-to-end fault tolerance tricky. As one concrete example, **Figure 2** shows a real bug case that occurred in a

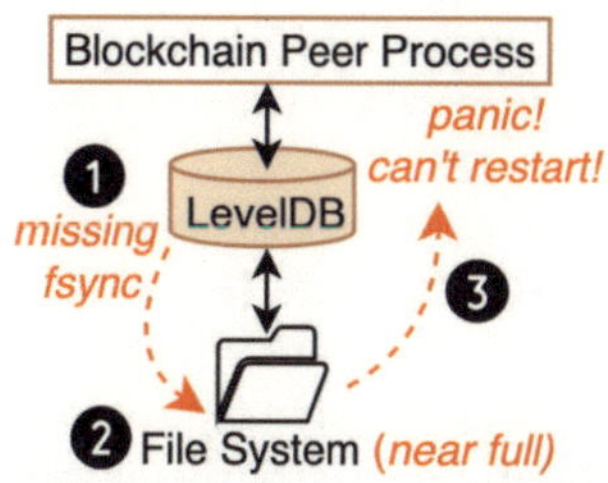

Figure 2.
A cross-layer issue in blockchain storage. The peer process panicked and failed to restart on a file system.

Blockchain storage system (Hyperledger Fabric [28]), where a peer process panicked and failed to restart [29]. While the root cause lies in the state DB (i.e., LevelDB), which misses a `fsync` syscall, the manifestation of the problem depends on the state of the underlying file system (i.e., near full). The issue was labeled with the *Highest* priority but it took 5 months to resolve due to the complexity [29]. Similar issues caused by cross-layer dependencies have led to widespread failures of major Blockchains in practice [30–34].

2.2 Representative approaches

Addressing the grand challenges require research innovations along multiple dimensions. In particular, *Testing* and *Debugging* are two complementary and essential techniques for ensuring the fault tolerance of data storage systems in general.

Specifically, *Testing* is a proactive approach used to identify potential defects in target storage systems before they impact a production environment. This method involves simulating various workload conditions, injecting faults, and stress-testing the system to evaluate its robustness under different failure scenarios. Techniques such as stress testing, fault injection, and endurance testing help uncover weaknesses in storage hardware, software, and network configurations. By running extensive test cases, system testing can detect performance bottlenecks, hardware limitations, and software bugs early in the development cycle, reducing the risk of unexpected failures. However, despite its effectiveness, system testing cannot anticipate every possible failure, especially those arising from complex real-world interactions between system layers.

On the other hand, *Debugging* is a reactive process aimed at diagnosing and resolving failures that have already occurred in a deployed system. This method involves analyzing system logs, performance metrics, and memory dumps to trace the root cause of an issue. Debugging tools can reconstruct system states, perform comparative analysis with functional systems, and detect anomalies using advanced diagnostic techniques. Common debugging challenges include identifying failures caused by rare concurrency issues, software-hardware interactions, or hidden firmware bugs. Accurate root cause identification is critical, as misdiagnosis can lead to ineffective fixes, prolonged system downtime, and unnecessary hardware replacements. As storage systems become increasingly complex, debugging requires sophisticated tools and expertise to ensure timely and effective failure resolution.

In the rest of this section, we first introduce a few representative techniques for analyzing the fault tolerance of individual data storage systems (Sections 2.3–2.5). We classify the works based on the layers of the general architecture of storage systems (**Figure 1**). Next, as one step toward addressing the open challenge, we discuss potential solutions for analyzing the fault tolerance of data storage systems in an end-to-end manner (Section 2.6).

2.3 Analyzing storage device

As discussed in Section 1, storage devices (e.g., HDD, SSD, and PM) serve as the foundation of data storage systems. They are built with different technologies and exhibit different characteristics, which may lead to different failure modes and require careful design of fault tolerance mechanisms at the upper layer. To understand the device-level behaviors, many researchers have conducted experiments and analyses on various devices at different granularity, including HDDs (e.g., [35, 36]),

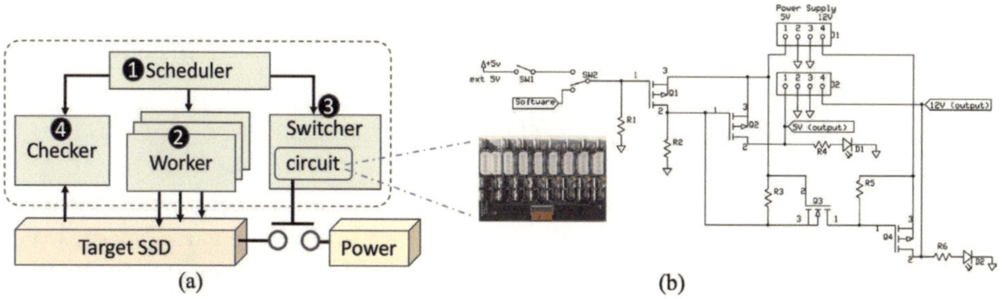

Figure 3.
The power fault injection framework for analyzing SSDs. (a) Workflow and (b) Fault injection circuit. (Adapted from: Ref. [17]).

HDD-based RAID systems (e.g., [37, 38]), raw flash memory chips (e.g., [39–48], flash-based SSDs (e.g., [49–51]), and PMs (e.g., [5, 52, 53]). We briefly describe one representative work below and refer the interested readers to [51] for more details.

Figure 3 shows one example of fault injection technique at the device layer [51]. As shown in **Figure 3a**, the framework includes four major components: *Scheduler*, *Worker*, *Switcher*, and *Checker*. The Scheduler coordinates the whole testing procedure including initializing the target SSD, selecting Workers to apply carefully designed workloads to the target for a predefined period, sending a signal to the Switcher at a random time within the working period, which turns off the power supply to the SSD accordingly, and invoking the Checker to read the special records present on the restarted SSD and check the correctness of the device state based on the record format. The Switcher component includes a dedicated circuit (**Figure 3b**) to enable efficient power fault injection with high fidelity. The fault injection testing procedure is executed iteratively and all issues found are written to logs for postmortem analysis. The framework has been applied to analyze dozens of SSDs from different vendors and exposed multiple failure modes of SSDs (e.g., shorn writes, serialization errors, bit corruption, and metadata corruption) that are different from traditional HDDs. The unique failure modes revealed in the experiments suggest the need for hardware awareness in building fault-tolerant storage systems. For example, serialization errors exposed by the framework imply that traditional fault-tolerance solutions relying on the correct order of operations (e.g., write-ahead logging in databases or journaling file systems) might not be sufficient. Similarly, metadata corruptions and shorn writes imply that an update-in-place to a sole copy of data is not enough to ensure fault tolerance. These findings have raised awareness and interest of power loss protection and relevant fault tolerance issues (e.g., crash consistency) in general, and the work has inspired follow-up research on reliability across the storage stack in the community (e.g., [54, 55]).

2.4 Analyzing local storage software stack

The local storage software stack serves as a critical intermediary between physical storage devices and user-level applications to manage data on a single computer (Section 1). Due to the prime importance, great efforts have been made to analyze and/or improve the reliability of the local storage software, including Linux fault injection infrastructure [56], regression test suites (e.g., xfstest [57] and e2fsprogs [58]), configuration dependency analyzers (e.g., CONFD [59]), fuzzers (e.g., Syzkaller [60]), and record-replay tools for simulating faults and testing the crash consistency of storage software (e.g., [61–63]). We briefly describe two representative techniques below and refer the interested readers to [56, 64] for more details.

2.4.1 Linux fault injection infrastructure (LFI)

This is a fault injection framework introduced to the Linux kernel since version 2.6.20 [56]. LFI can inject faults into the Linux kernel to simulate various issues (i.e., memory access errors) [65]. The implementation of this feature is located in the Linux source code under a dedicated path (`lib/fault-inject.c`), and the LFI capabilities can be configured either at boot time or during runtime [56, 65]. Roughly speaking, the LFI module reads the input parameters (e.g., interval and probability) from the `debugfs` string and stores the result in the fault attribute structure (`fault_attr`). The core function (`should_fail_ex`) examines the capabilities, returns true if they are met, indicating a failure, or returns false if not. Note that the LFI feature is turned off by default, and activating it requires declaring specific directives in the kernel configuration and recompilation.

The LFI contains various capabilities targeting different components in the Linux kernel [56]. For example, `failslab` allows injecting slab allocation failures in the kernel memory allocation functions (e.g., `kmalloc()` and `kmem_cache_alloc()`) [66]. Similarly, `fail_page_alloc` allows injecting failures in memory page allocation, which can affect all the functions related to paging (e.g., `alloc_pages()` and `get_fee_pages()`).

Figure 4 illustrates an example workflow of LFI to inject faults at a device driver. After configuring and compiling the kernel for fault injection, the next step is injecting faults into the target device through the `debugfs` interface [67]. The fault injection is done by filling the fault capabilities directly through `debugfs`, where the target device driver that has been designed to be injected reads those capabilities and executes the injection action. LFI includes a predefined script (`failcmd.sh`) to inject the fault, which is basically a bash script that utilizes `debugfs`. When the injection value is received in the target device driver, the function (`should_fail`) is triggered and causes the failure. Since LFI is at the OS level, errors can be directed to the user level for further analysis. Finally, the `dmesg` tool collects kernel messages, including

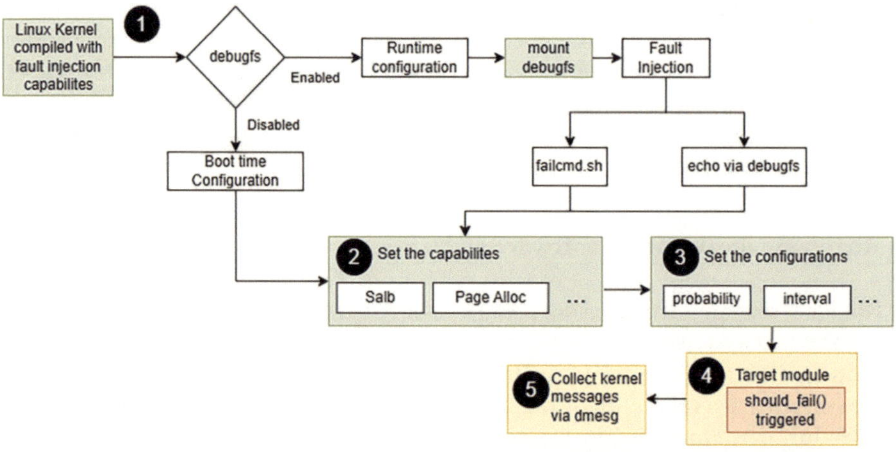

Figure 4.
An example workflow of Linux fault injection.

any errors caused by the `should_fail` function. Note that LFI is designed to be extendable to include new capabilities. It can also support additional fault models by adding the `should_fail` function to different locations in the kernel source code.

2.4.2 The RFSCK framework

This is one research prototype for injecting faults and analyzing the resilience of local file system checkers [64]. As shown in **Figure 5**, there are ten main steps to test the fault tolerance of file system checkers systematically *via* RFSCK: (1) the framework makes a copy of the test image which contains a corrupted file system; (2) the target checker (i.e., `fsck`) is executed to check and repair the original corruption on the copy of the test image; (3) after `fsck` finishes normally in the previous step, the resulting image is stored as the *reference image*; (4) during the checking and repairing of `fsck`, the fault injection tool `rfsck-test` records the I/O commands generated by `fsck` in a command history file (the basic mode); (5) the framework makes another copy of the original test image; (6) `rfsck-test` replays partial commands recorded in step 4 to the new copy of the test image, which emulates the effect of an interrupted `fsck`; (7) the image generated in step 6 is stored as the *interrupted image*; (8) `fsck` is executed again on the interrupted image to fix any repairable issues; (9) the image generated in step 8 is stored as the *repaired image*; and (10) finally, the framework compares the file system on the repaired image with that on the reference image to identify any mismatches. Note that in step 8, `fsck` has been executed without interruption, so a mismatch implies that there is some corruption that cannot be recovered by `fsck`. Also, besides the basic mode shown in the figure, RFSCK includes an advanced mode for testing file system checkers with logging support [64]. As of this writing, the RFSCK framework has been applied to test the checkers of multiple widely used file systems (i.e., `e2fsck` [58] for Ext-series file systems, `xfs-repair` [57] for XFS file system, `btrfs-fsck` for BtrFS file system, and `f2fs-fsck` for F2FS file system). The experimental results have demonstrated multiple vulnerabilities in the local file system layer (e.g., the file system may be left in an uncorrectable state if the repair is interrupted [68]).

2.5 Analyzing large-scale storage systems

Enabled by local storage software stack, large-scale storage systems manage resources across computer nodes to support various services and application scenarios at scale (e.g., cloud object storage, distributed file systems, and blockchains) [15–23, 28, 69–73]. Similar to other practical software systems, large-scale storage systems

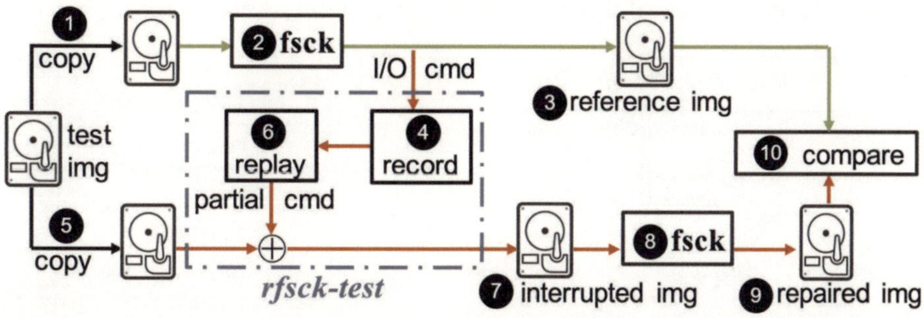

Figure 5.
The RFSCK framework for analyzing the fault tolerance of file system checkers. (Adapted from: Ref. [18]).

typically have built-in test suites to ensure their robustness (e.g., [15, 19]). In addition, many research prototypes have been proposed to understand the reliability of distributed systems (e.g., [74–85]). We briefly describe two representative frameworks below and refer the interested readers to [86, 87] for more details.

2.5.1 The PFAULT framework

As one example of analyzing the fault tolerance of large-scale storage systems, we introduce a scalable fault injection framework called PFAULT, which has been used to analyze the failure handling and recovery of multiple storage systems at scale [87–89]. As shown in **Figure 6**, PFAULT includes four major components: *Failure State Emulator*, *PFS Worker*, *PFS Checker*, and *Orchestrator*. And it can be connected to the target distributed/parallel storage system (which may include different types of nodes as described in Section 1) *via* remote storage protocols (e.g., iSCSI protocol [90]). More specifically, the *Failure State Emulator* is the key component responsible for injecting faults to trigger the fault tolerance procedures of the target system. It mounts a set of virtual devices to the storage nodes *via* iSCSI and forwards all device I/O commands to the backing files through the protocol, each of which represents the persistent state of a corresponding storage node. Moreover, the persistent states of storage nodes are manipulated by *Failure State Emulator* to emulate system failure states based on workloads and a set of predefined fault models (e.g., *Whole Device Failure*, *Global Inconsistency*, and *Network Partitioning*) derived from real-world failure scenarios reported in the literature [35, 36, 65, 91–94]. Besides the *Failure State Emulator*, the *PFS Worker* launches workloads to exercise the target system and generate I/O operations; the *PFS Checker* invokes the failure handling and recovery utilities (i.e., LFSCK for Luster) of the target system as well as a set of verifiable workloads to examine the fault tolerance of the target system thoroughly; and the *Orchestrator* component coordinates the overall workflow and collects the corresponding logs automatically for in-depth fault tolerance analysis. As of this writing, the PFAULT framework has been applied to study two large-scale production storage systems including Luster [19] and BeeGFS [15]. The experimental results have exposed multiple cases where the target systems' fault tolerance guarantee is imperfect (e.g., the recovery procedure itself may hang, fail abruptly, or trigger kernel panics when scanning the storage nodes in the cluster [88]).

2.5.2 AWS fault injection simulator (FIS) service

Since the publication of PFAULT [87], many efforts have been made to improve further the fault tolerance of distributed systems. Notably, Amazon recognizes the

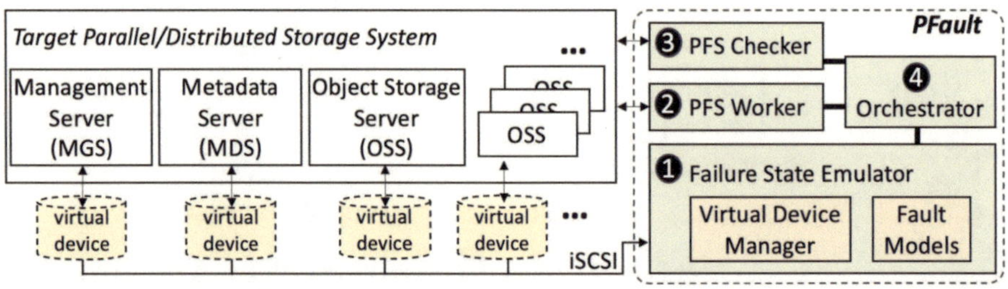

Figure 6.
The PFAULT framework for analyzing the fault tolerance of large-scale storage systems. (Adapted from: Ref. [18]).

importance of fault injection and commercializes a service called Fault Injection Simulator (FIS) [86]. The FIS shares similar design goals and principles as PFAULT but extends the target to general distributed systems. To achieve generality, it allows integration with third-party utility programs (e.g., stress-ng [95]) to conduct comprehensive testing and measurement. On the other hand, since FIS relies on utility programs running *inside* the target system to simulate faults, it might potentially change the target system and affect the fidelity. Therefore, FIS is more suitable for testing user-level applications instead of testing a full system stack. Also, FIS is a commercial service relying on other AWS services (e.g., EC2 and CloudWatch) to work, which might not be applicable to on-premise systems or non-commercial use cases with a limited budget. Therefore, additional efforts are probably needed to make the comprehensive fault tolerance analysis capability generally available.

2.6 Putting it all together: Toward full-stack fault tolerance analysis

The frameworks introduced in Sections 2.3–2.5 represent state-of-the-art techniques analyzing the fault tolerance of corresponding storage systems. While they are excellent for their original purposes, they are still insufficient to address the end-to-end fault tolerance challenge because they cannot capture the inherent dependencies across layers. To address the limitation, researchers have looked into full-stack approaches to improve the coverage. Notably, VINTER [61] is a framework to support full-system testing of PM-based storage systems. By leveraging virtual machine (VM) technology, VINTER is designed to host a complete software stack. VINTER has been applied to test PM-based file systems in the Linux kernel (e.g., PMFS [13]) and has helped expose multiple crash-consistency bugs that affect the fault tolerance of target systems [61].

Unfortunately, while the VINTER approach is promising, our experiments found that it is not scalable enough. There are multiple limitations based on our analysis: First of all, VINTER only supports minimal-built OS, which makes it incompatible with most application scenarios that depends on important system libraries (e.g., PMDK for PM application development). Second, VINTER incurs significant overhead even under a small set of writes from the applications (e.g., for a workload with 256*20 bytes of writes, VINTER cannot finish the testing within 12 hours), which suggests that VINTER cannot be applied to real-world scenarios where write operations are common and write sizes are typically larger than a few KBs. Third, VINTER has little support for debugging the fault tolerance issues triggered and pinpointing the root causes.

Inspired by VINTER as well as the key observations on its limitations, we propose a scalable VM-based framework called VFAULT for analyzing the fault tolerance of the entire storage system stack on PM. As shown in **Figure 7**, the framework leverages customized VMs to emulate storage devices (e.g., PM) and host the entire software stack from OS kernel to applications. It supports a set of critical features that are important for scalable fault tolerance analysis (e.g., record and replay, tracing and debugging, parallel crash state generation and testing). We elaborate on the main steps below.

The VFAULT workflow begins with an initial input system image that encompasses the full-system software stack (e.g., the OS kernel, applications, and required libraries, and all environmental configurations). The guest OS was initiated with root privileges in the VM. After booting the system, the framework enters the recording phase (i.e., "VM: Record"). It captures a snapshot as the original snapshot, comprising

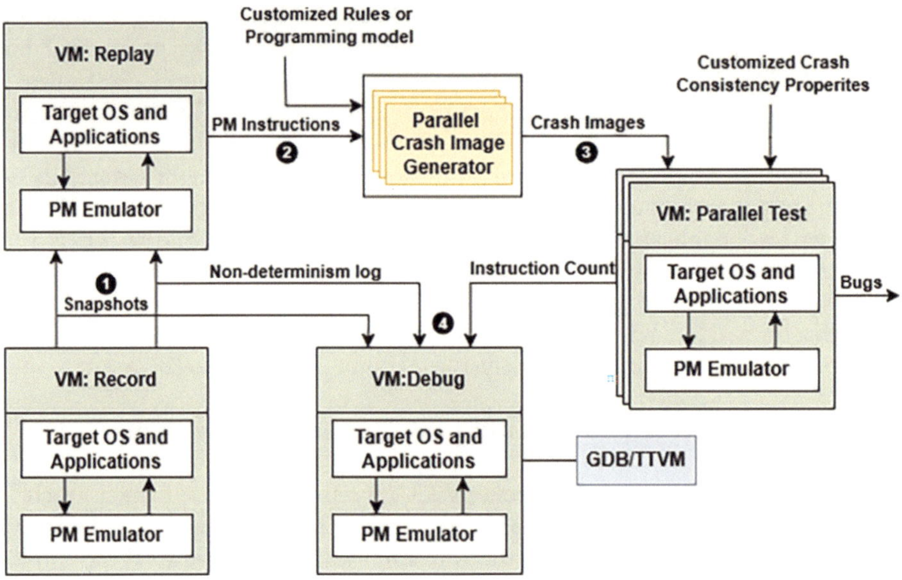

Figure 7.
A VM-based framework (VFAULT) for analyzing the full system stack on PM.

emulated device status (e.g., PM) and CPU register information. Moreover, the framework runs testing workloads on the pre-generated snapshot and leverages an architecture-neutral dynamic analysis tool called PANDA [96] to record the execution of the whole system stack under the testing workloads, which captures relevant system calls and critical device commands/instructions (e.g., PM instructions). The recording phase generates two types of files (i.e., *snapshots* and *non-determinism logs*) to be utilized for testing and debugging support in the following steps.

Next, in the replay phase (i.e., "VM: Replay"), the framework extracts storage commands and/or memory instructions (e.g., PM instructions) from the recorded executions. The framework generates a series of crash images based on the extracted instructions/commands and corresponding constraints (e.g., specifications of PM programming models and customized rules). These crash images represent the persistent states of the target systems under various fault scenarios, which can be used for parallel testing and fault tolerance analysis.

In the testing phase (i.e., "VM: Parallel Test), the framework checks the recovery of the target system on crash states, and validates if there are any crash consistency issues (i.e., system fails to recover to a consistent state from the crash images). Given the complexity of the storage system and the variety of possible crash states, the framework runs concurrent VMs to generate crash images in parallel (i.e., Parallel Crash Image Generator) and test the recoverability and crash consistency in parallel too. In this way, framework can improve the scalability with reasonable tradeoffs (e.g., VM resources vs. scalability). In addition, the framework is integrated with both classic debugging tools (e.g., GDB [97]) and time-traveling virtual machine (TTVM) [98] to support pinpointing the root causes of crash consistency issues exposed.

Note that the VFAULT framework is designed to be extensible to support third-party debugging tools, customized crash consistency rules, and future programming models. One key technique enabling these features is the snapshot mechanism. The framework generates a privileged snapshot immediately after the guest OS boots. This snapshot has two critical components: (1) Memory Status, which includes all data residing in the current emulated device; (2) CPU States, which includes the current register values for the emulated CPU. Subsequently, the snapshot is incorporated into various phases of the framework's execution. For example, in the recording and tracing phase, we load the pre-generated snapshot to support executing the workload to generate critical memory operations (e.g., fence and cache line flush instructions on PM). Similarly, in the testing phase, all subsequent crash states are generated based on the updates to the original snapshot (and thus creating additional snapshots for representing crash images).

At the time of this writing, a prototype of the VFAULT framework has been applied to test the crash consistency of full storage software stack under multiple application scenarios. **Table 1** shows one set of experimental results for testing PM-based systems. In this set of experiments, VFAULT emulated a 128 MB PM as the storage device and configured it accordingly through the guest OS kernel in VM. We evaluated six PM applications on top of the PM software stack, including *B-tree, C-tree, RB-tree, Hashmap_atomic, Hashmap_tx* and *Hashmap_rp* (as listed in the first column of **Table 1**). All PM applications relied on the PMDK library and Ext4-DAX file system support. Following the generation of crash states, we employed the Ext4 file system checker `e2fsck` to double check any corrupted states at the file system level. Moreover, we manually examined the recovered data structures of the workloads to validate any potential corruptions.

The experiments exposed multiple fault tolerance issues of the target PM system stack. As summarized in **Table 1**, we observed five different failure symptoms including checksum errors ("Cksum Err"), journal transaction corruptions ("J-Txn Err"), metadata corruptions ("Metadata Err"), unmount errors ("Umount Err"), and watchdog bugs reported within the guest OS ("Watchdog Bug"). Further analysis indicates that some fault tolerance issues may be caused by the interplay and dependency between PM library (PMDK) and file system components (`e2fsck`), which further suggests the importance of the holistic approach for ensuring end-to-end fault tolerance in practice. Also, the current prototype of VFAULT focuses on a single-node storage system stack; additional research and engineering efforts are needed to extend it to large-scale storage systems, which we leave as future work.

Applications on full system stack	Cksum Err	J-Txn Err	Metadata Err	Umount Err	Watchdog bug
B-tree	4	1	3	1	1
C-tree	29	4	1	1	1
RB-tree	17	2	1	0	1
Hashmap_atomic	8	4	1	3	1
Hashmap_tx	8	4	0	0	1
Hashmap_rp	6	4	0	0	0
TOTAL	72	19	6	5	5

Table 1.
Full-system testing results under six application scenarios.

3. Conclusion and future work

In this chapter, we have described the general architecture of data storage systems, which mainly includes three layers: storage device, local storage software stack in the OS kernel, and user-level applications, which may be distributed at scale. We have also discussed the design and implementation of multiple representative fault injection testing frameworks for individual storage systems. While these frameworks are excellent for their original design goals, targeting a single layer, they are relatively limited from an end-to-end perspective because there are inherent dependencies across layers that may affect the end-to-end fault tolerance guarantees of data storage systems in practice. Given the complexity of real-world data storage systems, we believe this is an open challenge that probably requires collective efforts from the communities. As one step toward addressing the grand challenge, we presented a VM-based full-stack testing framework called VFAULT, which currently focuses on the single-node storage system stack. Additional research efforts are likely needed to extend the idea of analyzing the fault tolerance of large-scale storage systems in an end-to-end manner, which we leave as future work. We hope that the comprehensive description of data storage systems and representative solutions presented in this chapter can inspire follow-up research on analyzing cross-layer dependencies and ensuring end-to-end fault tolerance for mission-critical data storage systems (e.g., distributed databases and blockchain storage) in general.

Acknowledgements

The authors would like to thank the members of the Data Storage Lab (DSL) for research discussions and experiments on multiple projects covered by this chapter, including Jinrui Cao, Om Rameshwar Gatla, Runzhou Han, Tabassum Mahmud, Chao Shi, Wei Xu, Varun S. Girimaji, Joshua John, Roop Kiran, and Vidhya Mannathu Parambil. This material was partially supported by the US National Science Foundation (NSF) under grants CNS-1566554/1855565 (CRII), CCF-1910747, CNS-1943204 (CAREER), CNS-2402858, and a Global Research Outreach (GRO) Award from Samsung Advanced Institute of Technology (SAIT) and Samsung Research America (SRA). Any opinions, findings, conclusions, or recommendations expressed in this material are those of the author(s) and do not necessarily reflect the views of the sponsors.

Author details

Mai Zheng*, Duo Zhang and Ahmed Dajani
Iowa State University, Ames, United States

*Address all correspondence to: mai@iastate.edu

References

[1] HotHardware: Windows 10 20H2 Update Reportedly Damages SSD File Systems If You Run ChkDsk. Available from: https://hothardware.com/news/windows-10-20h2-update-damages-ssd-file-systems-chkdsk

[2] When Solid State Drives are Not That Solid. Available from: https://blog.algolia.com/when-solid-state-drives-are-not-that-solid/ [Accessed: July 14, 2019]

[3] raid0: Data Corruption When Using Trim. Available from: https://www.spinics.net/lists/raid/msg49440.html [Accessed: July 19, 2015]

[4] Intel® Optane™ PMem. Available from: https://www.intel.com/content/www/us/en/products/details/memory-storage/optane-dc-persistent-memory.html

[5] Yang J, Kim J, Hoseinzadeh M, Izraelevitz J, Swanson S. An empirical guide to the behavior and use of scalable persistent memory. In: 18th USENIX Conference on File and Storage Technologies (FAST 20). Santa Clara, CA, USA: USENIX Association; 2020. pp. 169-182

[6] Patterson DA, Gibson G, Katz RH. A case for redundant arrays of inexpensive disks (RAID). In: Proceedings of the 1988 ACM SIGMOD International Conference on Management of Data, SIGMOD '88. New York, NY, USA: ACM; 1988. pp. 109-116

[7] Arpaci-Dusseau RH, Arpaci-Dusseau AC. Operating Systems: Three Easy Pieces; 2018

[8] Love R. Linux Kernel Development. United States of America: Pearson Education; 2010

[9] Nordvik R. Ext4. In: Mobile Forensics–The File Format Handbook: Common File Formats and File Systems Used in Mobile Devices. Cham, Switzerland: Springer; 2022. pp. 41-68

[10] Rodeh O, Bacik J, Mason C. Btrfs: The linux b-tree filesystem. ACM Transactions on Storage (TOS). 2013;**9** (3):1-32

[11] Lee C, Sim D, Hwang J, Cho S. {F2FS}: A new file system for flash storage. In: 13th USENIX Conference on File and Storage Technologies (FAST 15). Santa Clara, CA, USA: USENIX Association; 2015. pp. 273-286

[12] Xu J, Swanson S. NOVA: A log-structured file system for hybrid volatile/non-volatile Main memories. In: Proceedings of the 14th USENIX Conference of File and Storage Technologies (FAST). Santa Clara, CA, USA: USENIX Association; 2016

[13] Dulloor SR, Kumar S, Keshavamurthy A, Lantz P, Reddy D, Sankaran R, et al. System software for persistent memory. In: Proceedings of the 9th European Conference on Computer Systems (EuroSys). New York, NY, United States: Association for Computing Machinery; 2014

[14] POSIX.1-2017 Specifications. Available from: https://pubs.opengroup.org/onlinepubs/9699919799/ [Accessed: July 14, 2019]

[15] BeeGFS File System. Available from: https://www.beegfs.io/

[16] Colossus. Available from: https://cloud.google.com/blog/products/storage-data-transfer/a-peek-behind-colossus-googles-file-system

[17] DAOS. Available from: https://ethereum.org/en/dao/

[18] Gluster FS. Available from: https://www.gluster.org

[19] Lustre File System. Available from: http://lustre.org/

[20] RedHat. Data Services: Red Hat Ceph Storage. Available from: https://www.redhat.com/en/technologies/storage/ceph

[21] Swift O. Available from: https://wiki.openstack.org/wiki/Swift

[22] The OrangeFS Project. Available from: http://www.orangefs.org/

[23] WekaIO. Available from: https://www.weka.io

[24] Top500 Supercomputers. Available from: https://www.top500.org/lists/top500/2019/11/

[25] LFSCK: An Online File System Checker for Lustre. Available from: https://github.com/Xyratex/lustre-stable/blob/master/Documentation/lfsck.txt

[26] I. I. S. A. Library. Available from: https://www.intel.com/content/www/us/en/developer/tools/isa-l/overview.html [Accessed: April, 2024]

[27] Reed IS, Solomon G. Polynomial codes over certain finite fields. Journal of the Society for Industrial and Applied Mathematics. 1960;**8**:300-304

[28] Androulaki E, Barger A, Bortnikov V, Cachin C, Christidis K, De Caro A, et al. Hyperledger fabric: A distributed operating system for permissioned blockchains. In: Proceedings of the Thirteenth EuroSys Conference (EuroSys). New York, NY, United States: Association for Computing Machinery; 2018

[29] Hyperledger Fabric. Hyperledger Bug: Full Disks Causes Level DB Corruption. 2021. Available from: https://jira.hyperledger.org/browse/FAB-18304

[30] Blockchain.com. Incident Report for Blockchain: TRON Withdrawal Processing Delays due to Node Provider's Configuration Issues. 2022. Available from: https://status.blockchain.com/incidents/tcx1m4jcxy1p

[31] Jordan Pearson. How decentralized is decentralized finance? Amazon's Server Outage Took Down a 'Decentralized' Crypto Exchange. 2021. Available from: https://www.vice.com/en/article/wxdnxy/amazons-server-outage-took-down-a-decentralized-crypto-exchange

[32] Prathap M. Amazon Outage Causes Seven Hour Disruption at Crypto Exchanges, Raises Questions on 'decentralisation'. 2021. Available from: https://www.businessinsider.in/investment/news/aws-outage-shows-that-dexs-arent-are-decentralised-as-expected/articleshow/88186644.cms

[33] Dalton M. Amazon AWS Outage Affects Multiple Exchanges'. 2021. Available from: https://cryptobriefing.com/amazon-aws-outage-multiple-exchanges/

[34] Datta S. Gemini down Due to Amazon Web Services EBS Outage; Exchange Working on Restoring Functions. 2022. Available from: https://cryptoslate.com/gemini-down-due-to-amazon-web-services-ebs-outage-exchange-working-on-restoring-functions/

[35] Bairavasundaram LN, Goodson GR, Pasupathy S, Schindler J. An analysis of

latent sector errors in disk drives. In: Proceedings of the 2007 ACM SIGMETRICS International Conference on Measurement and Modeling of Computer Systems. New York, NY, United States: Association for Computing Machinery; 2007

[36] Schroeder B, Gibson GA. Disk failures in the real world: What does an MTTF of 1,000,000 hours mean to you? In: Proceedings of the 5th USENIX Conference on File and Storage Technologies (FAST'07). New York, NY, United States: Association for Computing Machinery; 2007

[37] Gibson G. Redundant Disk Arrays: Reliable Parallel Secondary Storage. PhD thesis. Berkeley: University of California; 1990

[38] Krioukov A, Bairavasundaram LN, Goodson GR, Srinivasan K, Thelen R, Arpaci-Dusseau AC, et al. Parity lost and parity regained. FAST. 2008;**2008**:127

[39] Belgal H, Righos N, Kalastirsky I, Peterson J, Shiner R, Mielke N. A new reliability model for post-cycling charge retention of flash memories. In: Proceedings of the 40th IEEE International Reliability Physics Symposium, IRPS'02. Dallas, TX, USA: IEEE; 2002

[40] Brand A, Wu K, Pan S, Chin D. Novel read disturb failure mechanism induced by flash cycling. In: Proceedings of the 31st IEEE International Reliability Physics Symposium, IRPS'93. Atlanta, GA, USA: IEEE; 1993

[41] Cai Y, Haratsch EF, Mutlu O, Mai K. Error patterns in MLC NAND flash memory: Measurement, characterization, and analysis. In: Proceedings of the Conference on Design, Automation and Test in Europe (DATE). Dresden, Germany: IEEE; 2012

[42] Cai Y, Yalcin G, Mutlu O, Haratsch EF, Unsal O, Cristal A, et al. Neighbor-cell assisted error correction for MLC NAND flash memories. In: The 2014 ACM International Conference on Measurement and Modeling of Computer Systems, SIGMETRICS'14. New York, NY, USA: ACM; 2014. pp. 491-504

[43] Grupp LM, Caulfield AM, Coburn J, Swanson S, Yaakobi E, Siegel PH, et al. Characterizing flash memory: Anomalies, observations, and applications. In: Proceedings of the 42nd Annual IEEE/ACM International Symposium on Microarchitecture (MICRO). New York, NY, USA: IEEE; 2009

[44] Grupp LM, Davis JD, Swanson S. The bleak future of NAND flash memory. In: Proceedings of the 10th USENIX Conference on File and Storage Technologies (FAST). San Jose, CA, USA: USENIX Association; 2012

[45] Kurata H, Otsuga K, Kotabe A, Kajiyama S, Osabe T, Sasago Y, et al. The impact of random telegraph signals on the scaling of multilevel flash memories. In: Symposium on VLSI Circuits, VLSI'06. Honolulu, HI, USA: IEEE; 2006

[46] Ong T, Frazio A, Mielke N, Pan S, Righos N, Atwood G, et al. Erratic erase in etox/sup tm/flash memory array. In: Symposium on VLSI Technology. Taipei, Taiwan: VLSI'93; 1993

[47] Suh K-D, Suh B-H, Lim Y-H, Kim J-K, Choi Y-J, Koh Y-N, et al. A 3.3V 32Mb NAND flash memory with incremental step pulse programming scheme. In: IEEE Journal of Solid-State Circuits. United States: JSSC'95; 1995

[48] Tseng H-W, Grupp LM, Swanson S. Understanding the impact of power loss on flash memory. In: Proceedings of the

48th Design Automation Conference (DAC'11). New York, NY, United States: Association for Computing Machinery; 2011

[49] Schroeder B, Lagisetty R, Merchant A. Flash reliability in production: The expected and the unexpected. In: Proceedings of the 14th USENIX Conference on File and Storage Technologies (FAST). Berkeley, CA, United States: USENIX Association; 2016

[50] Zheng M, Tucek J, Qin F, Lillibridge M. Understanding the robustness of SSDs under power fault. In: Proceedings of the 11th USENIX Conference on File and Storage Technologies (FAST). Berkeley, CA, United States: USENIX Association; 2013

[51] Zheng M, Tucek J, Qin F, Lillibridge M, Zhao BW, Yang ES. Reliability analysis of ssds under power fault. ACM Transactions on Computer Systems (TOCS). 2017;**34**:1-28

[52] Gatla OR, Zhang D, Xu W, Zheng M. Understanding persistent-memory-related issues in the linux kernel. ACM Transactions on Storage. 2023;**19**(4):1-28

[53] Zhang D, Gatla OR, Xu W, Zheng M. A study of persistent memory bugs in the linux kernel. In: Proceedings of the 14th ACM International Conference on Systems and Storage (SYSTOR). New York, NY, United States: Association for Computing Machinery; 2021

[54] Alter J, Xue J, Dimnaku A, Smirni E. Ssd failures in the field: Symptoms, causes, and prediction models. In: Proceedings of the International Conference for High Performance Computing, Networking, Storage and Analysis (SC). New York, NY, United States: Association for Computing Machinery; 2019. pp. 1-14

[55] Xu E, Zheng M, Qin F, Xu Y, Wu J. Lessons and actions: What we learned from 10k ssd-related storage system failures. In: Procedings of USENIX Annual Technical Conference (USENIX ATC). Berkeley, CA, United States: USENIX Association; 2019

[56] Linux Fault Injection Capabilities Infrastructure. Available from: https://docs.kernel.org/fault-injection/fault-injection.html

[57] XFS File System Utilities. Available from: https://access.redhat.com/documentation/en-US/Red_Hat_Enterprise_Linux/6/html/Storage_Administration_Guide/xfsothers.html [Accessed: July 14, 2019]

[58] e2fsprogs: ext2/3/4 Filesystems Utilities. Available from: http://e2fsprogs.sourceforge.net [Accessed: July 14, 2019]

[59] Mahmud T, Gatla OR, Zhang D, Love C, Bumann R, Zheng M. ConfD: Analyzing configuration dependencies of file Systems for fun and Profit. In: 21st USENIX Conference on File and Storage Technologies (FAST). Berkeley, CA, United States: USENIX Association; 2023

[60] Syzkaller – Kernel Fuzzer. Available from: https://github.com/google/syzkaller

[61] Kalbfleisch S, Werling L, Bellosa F. Vinter: Automatic non-volatile memory crash consistency testing for full systems. In: In 2022 USENIX Annual Technical Conference (USENIX ATC 22). Carlsbad, CA: USENIX Association; July 2022. pp. 933-950

[62] LeBlanc H, Pailoor S, KRE OS, Dillig I, Bornholt J, Chidambaram V. Chipmunk: Investigating crash-consistency in persistent-memory file systems. In: Proceedings of the

Eighteenth European Conference on Computer Systems. New York, NY, United States: Association for Computing Machinery; 2023. pp. 718-733

[63] Zheng M, Tucek J, Huang D, Qin F, Lillibridge M, Yang ES, et al. Torturing databases for fun and profit. In: Proceedings of the 11th USENIX Symposium on Operating Systems Design and Implementation (OSDI). Berkeley, CA, United States: USENIX Association; 2014

[64] Gatla OR, Hameed M, Zheng M, Dubeyko V, Manzanares A, Blagojević F, et al. Towards Robust File System Checkers. In: 16th USENIX Conference on File and Storage Technologies (FAST). New York, NY, United States: Association for Computing Machinery; 2018

[65] Subramanian S, Zhang Y, Vaidyanathan R, Gunawi HS, Arpaci-Dusseau AC, Arpaci-Dusseau RH, et al. Impact of disk corruption on open-source DBMS. In: IEEE 26th International Conference on Data Engineering (ICDE'10). Long Beach, CA, USA: IEEE; 2010

[66] Bonwick J et al. The slab allocator: An object-caching kernel memory allocator. In: USENIX summer, volume 16. Boston, MA, USA, Berkeley, CA, United States: USENIX Association; 1994

[67] Debugfs Tool. Available from: http://man7.org/linux/man-pages/man8/debugfs.8.html [Accessed: July 14, 2019]

[68] Gatla OR, Zheng M, Hameed M, Dubeyko V, Manzanares A, Blagojevic F, et al. Towards robust file system checkers. ACM Transactions on Storage (TOS). 2018;**14**:1-25

[69] Apache Hadoop. Available from: https://hadoop.apache.org/docs/stable/

[70] Apache Hadoop YARN. Available from: https://hadoop.apache.org/docs/current/hadoop-yarn/hadoop-yarn-site/YARN.html

[71] Apache Zookeeper. Available from: https://zookeeper.apache.org

[72] Ghemawat S, Gobioff H, Leung S-T. The Google file system. In: Proceedings of the 19th ACM Symposium on Operating Systems Principles (SOSP). New York, NY, United States: Association for Computing Machinery; 2003

[73] Lakshman A, Malik P. Cassandra: A decentralized structured storage system. ACM SIGOPS Operating Systems Review. 2010;**44**:35-40

[74] Alagappan R, Ganesan A, Lee E, Albarghouthi A, Chidambaram V, Arpaci-Dusseau AC, et al. Protocol-aware recovery for consensus-based storage. In: 16th {USENIX} Conference on File and Storage Technologies ({FAST} 18). New York, NY, United States: Association for Computing Machinery; 2018. pp. 15-32

[75] Barton JH, Czeck EW, Segall ZZ, Siewiorek DP. Fault injection experiments using fiat. IEEE Transactions on Computers. 1990;**39**(4): 575-582

[76] Dawson S, Jahanian F, Mitton T. Orchestra: A probing and fault injection environment for testing protocol implementations. In: Proceedings of IEEE International Computer Performance and Dependability Symposium (IPDS'96). Urbana-Champaign, IL, USA: IEEE; 1996

[77] Ganesan A, Alagappan R, Arpaci-Dusseau AC, Arpaci-Dusseau RH. Redundancy does not imply fault tolerance: Analysis of distributed storage

reactions to single errors and corruptions. In: 15th USENIX Conference on File and Storage Technologies (FAST). Berkeley, CA, United States: USENIX Association; 2017

[78] Gunawi HS, Do T, Joshi P, Alvaro P, Hellerstein JM, Arpaci-Dusseau AC, et al. Fate and destini: A framework for cloud recovery testing. In: Proceedings of the 8th USENIX Symposium on Networked Systems Design and Implementation (NSDI'11). Berkeley, CA, United States: USENIX Association; 2011

[79] Jepsen. Available from: https://github.com/jepsen-io/jepsen

[80] Leesatapornwongsa T, Hao M, Joshi P, Lukman JF, Gunawi HS. Samc: Semantic-aware model checking for fast discovery of deep bugs in cloud systems. In: 11th USENIX Symposium on Operating Systems Design and Implementation (OSDI'14). Berkeley, CA, United States: USENIX Association; 2014

[81] Lu J, Liu C, Li L, Feng X, Tan F, Yang J, et al. Crashtuner: Detecting crash-recovery bugs in cloud systems via meta-info analysis. In: Proceedings of the 27th ACM Symposium on Operating Systems Principles (SOSP'19). New York, NY, United States: Association for Computing Machinery; 2019

[82] Seungjae Han KG, Shin, Rosenberg HA. Doctor: An integrated software fault injection environment for distributed real-time systems. In: Proceedings of 1995 IEEE International Computer Performance and Dependability Symposium (IPDS'95). NW Washington, DC, United States: IEEE Computer Society; 1995

[83] Stott DT, Floering B, Burke D, Kalbarczpk Z, Iyer RK. Nftape: A framework for assessing dependability in distributed systems with lightweight fault injectors. In: Proceedings IEEE International Computer Performance and Dependability Symposium (IPDS'00). NW Washington, DC, United States: IEEE Computer Society; 2000

[84] Stuardo CA, Leesatapornwongsa T, Suminto RO, Ke H, Lukman JF, Chuang W-C, et al. Scalecheck: A single-machine approach for discovering scalability bugs in large distributed systems. In: 17th USENIX Conference on File and Storage Technologies (FAST'19). Berkeley, CA, United States: USENIX Association; 2019

[85] Yuan X, Yang J. Effective concurrency testing for distributed systems. In: Proceedings of the Twenty-Fifth International Conference on Architectural Support for Programming Languages and Operating Systems (ASPLOS'20). New York, NY, United States: Association for Computing Machinery; 2020

[86] Amazon. AWS Fault Injection Simulator (AWS FIS). 2023. Available from: https://docs.aws.amazon.com/fis/

[87] Cao J, Gatla OR, Zheng M, Dai D, Eswarappa V, Mu Y, et al. PFault: A general framework for analyzing the reliability of high-performance parallel file systems. In: Proceedings of the 2018 International Conference on Supercomputing (ICS). New York, NY, United States: Association for Computing Machinery; 2018

[88] Han R, Gatla OR, Zheng M, Cao J, Zhang D, Dai D, et al. A study of failure recovery and logging of high-performance parallel file systems. ACM Transactions on Storage (TOS). 2022;**18**(2):1-44

[89] Han R, Zhang D, Zheng M. Fingerprinting the checker policies of

parallel file systems. In: 2020 IEEE/ACM Fifth International Parallel Data Systems Workshop (PDSW). GA, USA: IEEE; 2020. pp. 46-51

[90] Meth KZ, Satran J. Design of the iscsi protocol. In: Proceedings of 20th IEEE/11th NASA Goddard Conference on Mass Storage Systems and Technologies (MSST). San Diego, California: IEEE; 2003

[91] Bairavasundaram LN, Arpaci-Dusseau AC, Arpaci-Dusseau RH, Goodson GR, Schroeder B. An analysis of data corruption in the storage stack. ACM Transactions on Storage. 2008;**4** (3):1, 28-8

[92] Ma A, Douglis F, Lu G, Sawyer D, Chandra S, Hsu W. Raidshield: Characterizing, monitoring, and proactively protecting against disk failures. In: 13th USENIX Conference on File and Storage Technologies (FAST'15). New York, NY, United States: Association for Computing Machinery; 2015

[93] Davidson SB, Garcia-Molina H, Skeen D. Consistency in a partitioned network: A survey. ACM Computing Surveys (CSUR). 1985;**17**(3):341-370

[94] Nightingale EB, Douceur JR, Orgovan V. Cycles, cells and platters: An empirical analysis of hardware failures on a million consumer PCs. In: Proceedings of the Sixth Conference on Computer Systems (EuroSys'11). New York, NY, United States: Association for Computing Machinery; 2011

[95] Colin Ian King. stress-ng: A Tool to Load and Stress A Computer System. Available from: https://wiki.ubuntu.com/Kernel/Reference/stress-ng

[96] Platform for Architecture-Neutral Dynamic Analysis. Available from: https://panda.re/

[97] Free Software Foundation. gdb: The GNU Debugger. Available from: https://man7.org/linux/man-pages/man1/gdb.1.html

[98] King ST, Dunlap GW, Chen PM. Debugging operating systems with time-traveling virtual machines. In: Proceedings of the 2005 USENIX Technical Conference. Berkeley, CA, United States: USENIX Association; 2005. pp. 1-15

Chapter 4

The Anatomy of Fault Diagnostics: Dissecting Definitions, Classifications, and Impact

Shahamat Shahzad Khan

Abstract

Electronic systems and components continuously evolve to become more accurate, efficient, dense, and reliable. These features have enabled the proliferation of electronic systems in all aspects of modern life, from consumer electronics to industrial applications. Continuous innovation has enabled us to use these electronic systems in some of the most critical applications, including surgical and medical equipment, control systems of electric vehicles, power plant cooling, monitoring and control systems, and aerospace navigation systems. The criticality of these applications demands the highest levels of reliability from these electronic systems and requires multiple reliability enforcement measures. Out of several reliability enforcement measures, fault tolerance (FT) is the most efficient and affordable solution. Efforts have been made to enhance electronic systems' reliability, primarily by introducing fault diagnostics (FDT) schemes and enforcing fault-tolerant (FT) designs. While FT allows the electronic system to continue to operate without interruption upon failure of its components, FDT detects and isolates the fault, making it the initial and the most crucial step in any FT strategy. The main objective of any FDT mechanism is to accurately detect and isolate or pinpoint the fault within the minimum possible time. Therefore, fault detection (FD) and fault isolation (FI) become the two main components of any FDT scheme. The importance of these two steps cannot be overemphasized because they trigger and guide the FT mechanism to take the appropriate action, thus ensuring the continuity of operation of electronic systems in critical applications.

Keywords: fault diagnostics, fault tolerance, reliability, fault detection, fault isolation

1. Introduction

Almost all engineering systems depend on electronic sub-systems that monitor, control, and power them. In the case of critical engineering systems, these electronic sub-systems must ensure a high availability factor of 99.999% [1] to help achieve the required reliability standards. Therefore, the availability of these electronic sub-systems is crucial to the system's overall functionality. These electronic systems are realized using several components, including but not limited to semiconductor switches, aluminum electrolytic capacitors (AEC), magnetic components such as transformers and inductors, and integrated circuits. The failure mode analysis of a particular group of electronic systems revealed that the semiconductor power devices are one of the most fragile components, second only to AECs, resulting in overall system shutdown and loss of service [2]. The same analysis can be applied to almost all electronic systems regardless of functionality.

During World War II, serious attempts were initiated to quantify and measure the reliability of electronic components. At that time, electronic tubes were the most failure-prone components in radars and radios. Soon, it was realized that the reliability of these components needed to be estimated and verified before their full-scale production. As a result, a method was developed to estimate these failure-prone components' reliability and life expectancy, which was the inception of reliability prediction and failure analysis for electronic components. Reliability prediction and environmental testing later became full-blown professional disciplines. Studies have shown that many factors can lead to unexpected failures in electronic systems. They include poor design, selection of low-quality components, manufacturing faults, design mistakes, or simple oversight.

Enforcing high-efficiency standards within a small footprint has strained the performance of high-availability systems by putting tremendous pressure on the performance and reliability of electronic systems and their components. For example, a high-speed processor requires fast delivery of large input currents in a very short duration, which poses the risk of controller failure. Similarly, the restrictive airflow due to high component densities and high operating temperatures in telecom switches and data center applications has posed additional challenges to the reliable operation of electronic systems and their power modules. Moreover, to achieve high power density and maximum efficiency, electronic systems and power modules usually consist of various component technologies, further straining the limits of the system design. Some of the factors that lead to an early system failure are:

1. Transient stress: Switching power devices often experience high transient stress, especially during turn-on and turn-off, that exceeds their safe operating area and causes a device failure.

2. Voltage overshoot: Device failure due to output voltage overshoot because of poor over-voltage protection and bad output filter design.

3. Capacitor failure: System failure due to a selection of poor-quality tantalum, ceramic, or electrolytic capacitors.

4. Thermal runaway: Inefficient thermal design results in system shutdown due to the thermal runaway of semiconductor power devices.

Irrelevant to the cause of system failure, the main aim of an FT design is to allow the system to operate after a fault within one or more of its components. FT is a system's ability to continue its operation under a single- or multiple-component failure. The purpose is to prevent disruption in operations arising from internal failures. This ensures business continuity and availability of mission-critical applications. So, how does the FT system achieve its robustness and availability? By deploying backup components that systematically replace compromised components during a catastrophic failure. The FT mechanism can only be executed if the fault's presence, location, and type are known, and these can be identified using an FDT mechanism. The initial step of the FDT mechanism is FD. FD determines the presence of a fault and time stamps it. It then triggers the FI mechanism, which helps identify the exact location of the fault and its type. Collectively, these two are termed fault detection and isolation (FDI) [3]. Once FDI is achieved, an FT mechanism is initiated to avoid system failure by severing the faulty component from the system and then re-configuring or replacing it with a redundant counterpart.

Now that it is established that FDT is crucial to implementing an FT design, we will look into the steps needed to develop an FDT mechanism. The first step in realizing an FDT mechanism is the selection of an FD signature. FD signatures are typically current and voltage signals from various critical locations of electronic systems. These signatures behave normally during the regular operation of the system. However, under a fault condition, the behavior of these signatures changes considerably to trigger an alarm. A better selection of FD signatures can help achieve swift detection and isolation of all types of faults. Selecting an FD signature defines the accuracy and effectiveness of an FDT mechanism and, ultimately, the FT mechanism. Care must also be taken to ensure that these signatures remain oblivious to transient conditions that are not faults but regular system operations. Several fault signatures are available, each with its benefits and shortcomings. Some signatures are fast but downgraded due to extra sensors' requirements or the inability to isolate a fault. At the same time, others may provide a more accurate analysis of the fault conditions at the expense of more considerable response time.

The main aim of an efficient FDT mechanism is to address the issues related to accuracy, cost, detection speed, and practical implementation. Modern engineering systems require an FDT mechanism that is cost-effective and accurate. Moreover, it should detect and pinpoint the fault within a few μs under varying operational conditions. Since FD signatures play a key role in the successful implementation of an FDT mechanism, it is better to lay down some of the important features of FD signatures.

1. Accuracy: This parameter requires that the FD signature accurately and consistently depict fault conditions, thus enabling the FDT algorithm to detect and isolate the fault. For instance, if the behavior of these FD signatures changes according to different fault types for the same components and remains consistent for a single fault under varying operating conditions, then the FDT algorithm can accurately differentiate between different fault conditions.

2. Extra sensor requirement: Adding additional sensors for measuring FD signatures adds to the manufacturing cost of the final product. A cost-effective FDT mechanism eliminates the need for additional sensors. By using already installed monitoring sensors and the system control signals for FDTs, the number of sensors installed can be reduced.

3. Multiple FD signatures: In large complex systems, FD and FI require more than one FD signature. This increases the overall cost of the FDT mechanism due to the requirement of multiple extra sensors. Moreover, output from multiple sensors needs higher processing resources in the central controller to decode the FD signatures. An FDT mechanism can be more effective if it uses a minimum number of FD signatures for multiple fault detection and isolation.

4. Detection speed: Critical engineering systems require swift FDI. With earlier information on the presence, location, and type of fault, the preceding FT

mechanism can take over immediately and prohibit the seepage of fault to the rest of the system.

5. Computational resource usage: In modern-day low-cost controllers used for implementing and monitoring the control scheme, the available computational resources, such as ALUs, ADCs, I/O ports, DAC, and so on, are often limited. Therefore, care should be taken to minimize the computational burden of processing the FD signatures.

6. Integration complexity: A minimum number of tapping points from the electronic systems should be used to extract these FD signatures. This allows for the quick integration of the FDT technique into existing engineering systems with no or fewer complications.

7. Engineering practicality: All practical systems will experience turbulence at their input and output. For example, load and source variation at the input and output ports is normal in electronic systems. An effective FD signature must be oblivious to the changes in these operating conditions. An FD signature should be robust to sudden variations in load and source. Moreover, the mechanism for specific operating conditions should be easily imported to any other frequency and power level.

2. Fault classification

A clear understanding of the common fault types in electronic systems and their classification is necessary to better understand the classification of various FD and FI techniques. This section briefly introduces the typical fault types in electronic systems and their root causes. Electronic device faults can be sectioned into three main groups and are also illustrated in **Figure 1**.

1. Catastrophic faults: These types of faults are responsible for a sudden service loss.

2. Parametric faults: Parametric or slow faults result from the gradual degradation of the component's performance due to aging.

3. Intermittent faults: These are sporadic and difficult to anticipate due to their complicated behavior.

Catastrophic faults in electronic modules mainly occur because of the failure of semiconductor devices due to short circuit faults (SCF) and open circuit faults (OCF). Refs. [4, 5] have discussed the specific causes of SCF, OCF, and component degradation. An SCF usually results from gate oxide shortening because of source metal shortening, gate metal shortening, and electric field stress. A SCF forces the switch to latch permanently irrespective of the value of the gate control signal and commonly results in high short circuit overcurrents. SCF introduces a sudden large inrush current, due to which these faults can be easily detected in an electronic system. These faults are quick and more damaging and require a swift FD and FI scheme. All electronic systems include a passive SCF protection scheme, such as quick blow fuses,

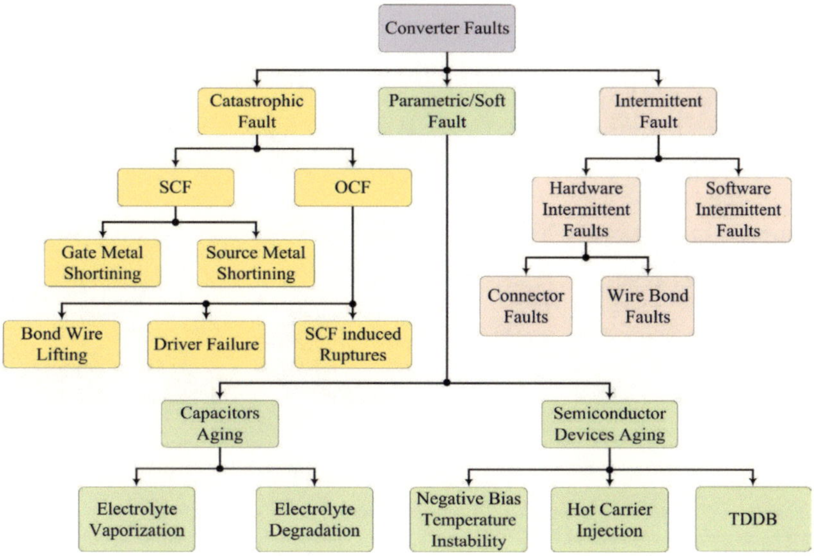

Figure 1.
Classification of typical faults in electronic systems.

or an active protection scheme, such as SCF detection circuits. OCF, on the other hand, results from a driver failure, SCF-induced rupture, and lifting of the bond wire during thermal cycling. The OCFs usually do not cause any sudden surge in current flow. They will disturb the system equilibrium regarding uneven magnetic flux distribution, which still causes power loss in a system. Due to the low severity of the OCF, their detection is often more complex than an SCF, and it requires more intricate FD and FI schemes. Regardless of the fault type, they must be detected and identified promptly. Otherwise, it will delay the initiation of the FT scheme, which can result in the penetration of the fault effects to normal components within the system. **Figure 2** shows several typical component damages due to SCF and OCF.

Parametric faults, also called soft faults, result from the aging of system components. AEC usually has the shortest life span than other electronic components such as semiconductors. Moreover, the semiconductor components also experience degradation due to phenomena such as negative bias temperature instability (NBTI), hot carrier injection (HCI), and time-dependent dielectric breakdown (TDDB). The aging of the AEC is the leading cause of the failure of DC-DC converters, which are the powerhouses of electronic systems. Vaporization and degradation of liquid electrolytes that occur due to the activation of the self-healing mechanism to recover the dielectric membrane in AEC are the main reasons that result in capacitor aging. Over-voltage stresses and high-temperature thermal shocks are the common causes of dielectric membrane damage, which triggers the self-healing mechanism. During this mechanism, the oxygen from the electrolyte reacts with the aluminum foil to produce AL_2O_3 and recover the dielectric membrane. This process also releases hydrogen gas. During the early life of the capacitor, the self-healing mechanism's activation rate is very low, resulting in a minute amount of hydrogen gas release. However, abnormal conditions, such as ambient temperature spikes, high RMS currents, voltage spikes, and high ripple currents, exponentially increase the frequency of the self-healing mechanism. This results in further vaporization of liquid electrolytes and the production of higher quantities of hydrogen gas, increasing the capacitor's internal pressure.

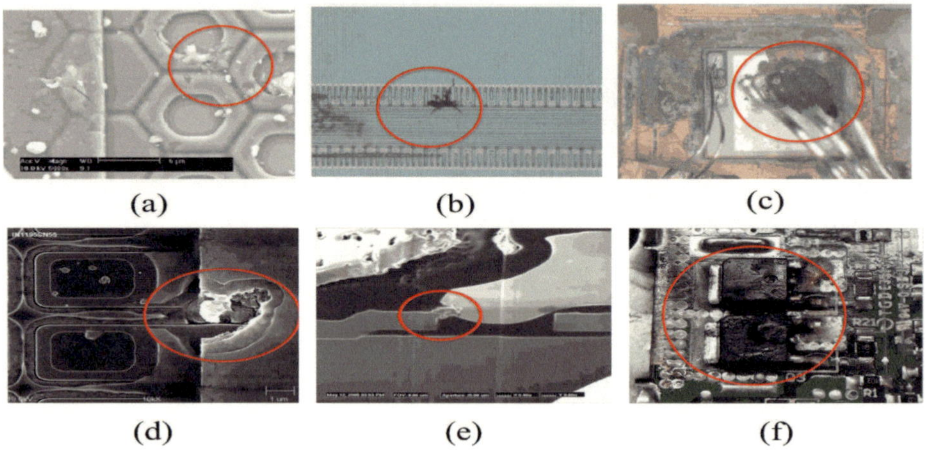

Figure 2.
Most common types of component failures: (a) Failure at the gate oxide of edge cell in BUK9508-55A [4] (b) Failure at the gate oxide of edge cell in BUK7Y3R0-40H [4] (c) Over-current burns at the center of die adjacent to wirebonds in BUK7L06-34ARC [4] (d) Source Metal to Gate Poly Short [5] (e) Gate Metal to Drain Poly Short [5] (f) MOSFET and the corresponding diver damage.

If unchecked, this pressure can increase to a point that causes the pressure release vent to open, resulting in a catastrophic failure of the AEC from drying out.

Intermittent faults, also known as the no-fault found, are highly concerning for engineers due to their irregular and unpredictable nature. The occurrence of these failures cannot be estimated nor replicated, and when a fault diagnostic routine is carried out, sometimes, no fault is found. In other words, these faults are hard to replicate and thus hard to find and ultimately troubleshoot. This category is broadly classified into hardware intermittent failure and software intermittent failure. Hardware intermittent failures result from connector and wire bond malfunction due to vibrations, stress relaxation, rapid environmental changes such as temperature fluctuations, and so on. Software intermittent failures occur when a particular set of conditions that are not standard operating procedures or unanticipated faulty working conditions are fulfilled. The most frequent failures in a software system are the unavailability of processing resources or storage memory due to some unanticipated operating conditions. As a result, the desired program can experience a crash due to limited resources, resulting in intermittent faults.

2.1 Fault handling mechanism

Continuous current and voltage stresses are some of the primary causes of failure. Electronic systems are designed to minimize these stresses on the internal critical components. However, interacting with various outside loads and sources, the systems and the components within experience high current and voltage transients. Naturally, external conditions such as abnormal temperature gradients, voltage surges, and current and load fluctuations are not ideal. As one of the remedial actions, component derating can improve the system life expectancy by utilizing components with higher-rated voltage or current specifications. By conservatively designing these systems, component derating can be used to increase the overall system reliability [6, 7], but it results in low power density and high costs [8]. A more pragmatic approach to improving the robustness and reliability of electronic systems was sought, which involved implementing FDT schemes and FT mechanisms. FDT is a wider term covering:

1. Fault detection: This is related to the study of detecting the presence and time of a fault.

2. Fault isolation: This is related to the study of identifying the location and type of the fault.

FDI is the primary step of an FT mechanism that detects the fault's presence and exact location [9]. Upon determining the precise fault location and type, the corresponding FT scheme is initiated either by removing/bypassing the faulty component [10, 11], leg/phase [12–14], or entire module [15] as shown in **Figure 3**.

As a specific case, the power electronic devices utilizing multilevel cascaded DC-DC converters for power delivery can ensure high reliability by implementing the system or subsystem redundancy. In case of a subsystem fault, the backup subsystem takes over, and the failed subsystem is bypassed as an FT measure [16–20]. This technique can avoid using detailed and complicated FDI schemes designed at the component level. However, as a downside to this technique, the overall implementation cost, weight, and volume get much more significant when a redundant subsystem is adjusted along with each cascaded subsystem to improve reliability.

The second approach works on a more detailed level and is called component-level redundancy. To implement it effectively with minimal expenses, the first step requires identifying the crucial components in the electronic systems more susceptible to failure. As a specific case of high-power DC-DC converters, the most failure-prone components are identified as the AECs and the high-power semiconductor components [2]. This study can be extended to almost all types of electronic systems. The next step involves the installation of a redundant switch, a leg, or a complete module in the power converter, which can replace the faulty component in case of failure in single-switch converters [10, 11, 21] or converters with a complex configuration [12–14, 22]. The component level redundancy method only requires the installation of redundant components for all the critical components rather than using a completely redundant system. This immensely reduces the implementation costs and space required to implement an FT system.

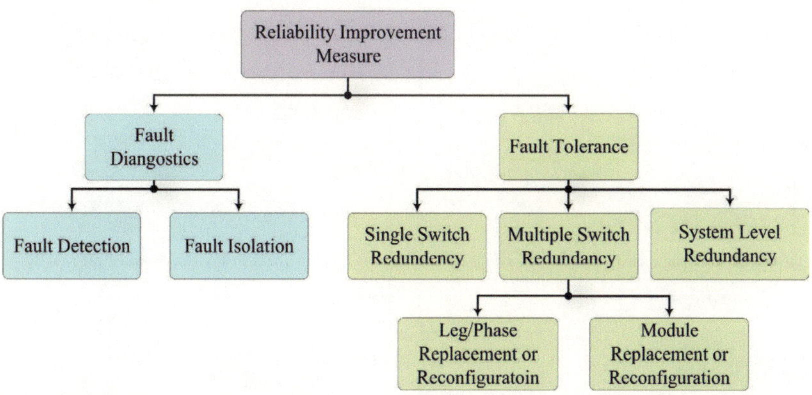

Figure 3.
Reliability improvement mechanism.

3. Fault diagnostic techniques classification

Figure 3 illustrates the reliability improvement schemes for implementing the FDT and FT. FDT is composed of two steps, FD and FI. In terms of implementation, the complete FDTs can be achieved using three main types of technologies: hardware-based, model-based, and historic data-based, which are illustrated in **Figure 4**.

3.1 Fault diagnostic signatures

Before diving into the classification of FDT techniques, getting the required information about the FDT signatures would be very helpful. Every electronic system has currents, voltages, or other specialized signals at various locations with a specific pattern during regular operation. This pattern will get distorted or its shape will deform when the system performs abnormally under fault. Termed as the FDT signatures, these signals get altered or affected due to the presence of any faults. These signals are extracted from critical circuit locations using various transducers or sensors. During regular operation, the pattern of FDT signatures is well defined; however, under abnormal or fault conditions, a deviation from the regular operating pattern due to a fault can help identify the faults within the electronic system.

These signals can be of any type, depending on the system. In most cases, these FDT signals are in the form of currents and voltages because of their inherent behavior of deviating quickly under abnormal conditions. However, recent developments in sensing technology and high signal processing speeds have encouraged researchers to use unorthodox signals to get the required information on the system's health. **Figure 5** lists some of the most effective FDT signatures used in electronic systems. It can be seen that apart from different types of current and voltage signals, temperature and electromagnetic interference (EMI) can also be used to measure the

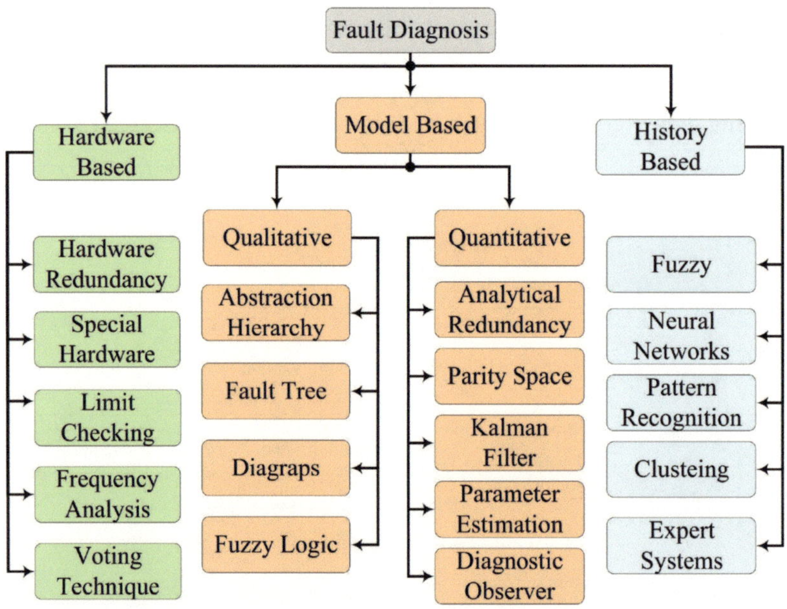

Figure 4.
Classification of FDT mechanisms.

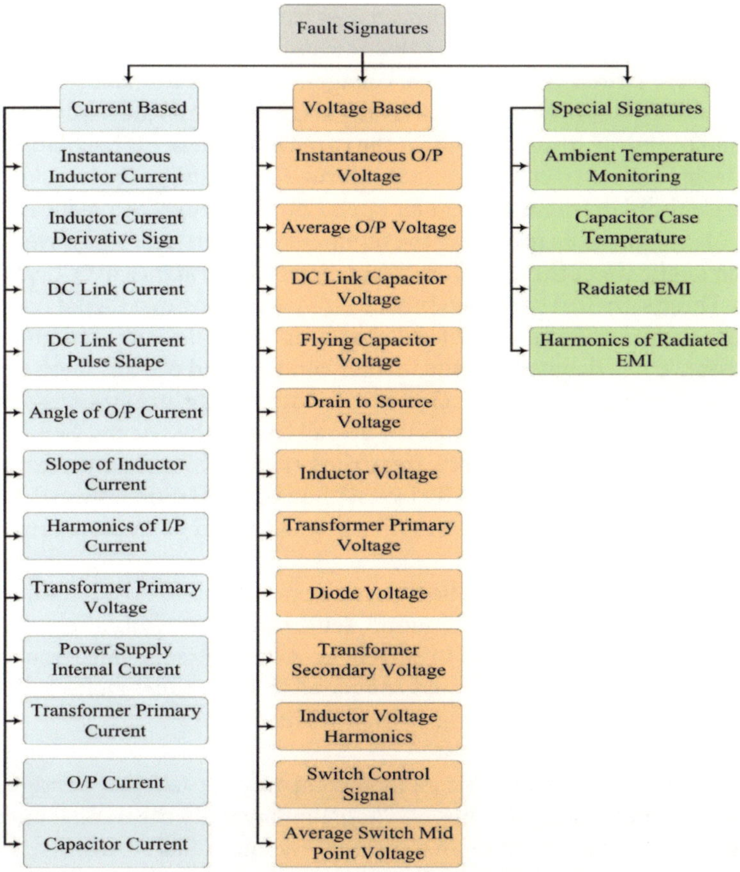

Figure 5.
Lists of FDT signatures.

system's health by detecting and isolating the fault. Once measured, these signatures can be used directly to establish a fact about the system's health, or more information can be extracted for better FDI in complex systems by calculating their derivatives, integrals, root mean square, or averages. In most cases, thresholds are defined as a limit for these FDT signatures, and when these limits are violated, a fault alarm is triggered.

3.2 Hardware-based FDTs

Hardware-based FDTs are used for SCF and OCF detection and isolation in electronic systems because they are swift and accurate. This approach uses analog or digital hardware to assess the pattern of the FDT signatures continuously. In most cases, a threshold defines the limit of these voltage or current signatures, and once this threshold is violated, the corresponding fault alarm is triggered. The hardware-based FDT can be classified into hardware redundancy, special hardware, limit checking, voting, and frequency analysis. These techniques can be utilized in specific circumstances. However, a combination of these techniques, working in unison, has often been used in a single FDT strategy. This can exponentially improve the overall effectiveness of the combined FDT schemes compared to a single FDT strategy.

1. Hardware redundancy: This approach uses hardware sensors to measure signals of interest or FDT signatures. Hardware redundancy commonly uses inductor currents, switch voltages, diode voltages, and capacitor voltages as the FDT signatures. This approach is simple, fast, and effective as the hardware sensors respond promptly to an anomaly in the patterns of these FDT signatures. Some FDT techniques extract the required information from the control signals used in the electronic systems for regular operation. This approach reduces the overall implementation costs regarding the number of sensors used. However, this approach might affect the overall control scheme performance in critical applications. These signatures are then processed by an analog or a digital diagnostics circuit or a controller such as a field programmable gate array or digital signal processor. These diagnostic units further compare and process these signatures to extract the required results. However, incorporating an extra sensor for FDT signature measurement increases the overall cost of the system. Moreover, this also introduces additional hardware to the system, which requires extra space and power. The designers usually frown upon this situation, especially in portable electronic devices where a PCB size, weight, and volume are critical.

2. Special hardware technique: The types of sensors used in this technique measure special signals such as pressure, sound, temperature, and vibrations. The outputs of these sensors are converted to current or voltage signals. Once transformed, these signals can be processed and compared to known thresholds to get the corresponding fault detection and isolation results.

3. Limit checking technique: This method compares the values of FD signatures with established predefined limits. These limits specify the normal operational range of the FD signatures of an electronic system. A fault alarm is triggered once the values of FDT signatures cross these predefined limits. The limit-checking technique requires defining the values of practical threshold limits that are usually small enough to get a rapid response in case of a fault but still robust to noise fluctuations and false triggers.

4. Frequency analysis technique: The frequency analysis technique is also very effective for fault detection and isolation. The frequency spectrum of FD signatures is consistent under normal operating conditions. However, new frequency components are added to the original spectrum under fault conditions. The frequency spectrum is studied under normal and fault conditions, and a specific spectrum is associated with certain types of faults to detect and isolate the exact fault location and type.

3.3 Model-based fault diagnostics

Hardware-based FDT mechanisms use sensors to measure FD signatures in the electronic systems, adding volume and cost to the overall system. This approach can bulk the overall system and prohibit its use in small and portable electronic devices. This shortcoming of hardware-based FDT mechanisms motivated researchers to opt for sensorless FDT mechanisms, also known as analytical redundancy, where a mathematical relationship is used to measure or estimate variables of the system model for anomaly detection. Based on a deep understanding of the system's physical properties, a mathematical model is developed that simulates the behavior of the actual system.

Initially, the mechanism compares the actual output to the model's predicted output to calculate the residual signal. These residuals are meant to remain negligible under no-fault conditions. However, they should become large enough to become noticeable under fault conditions. These residuals can be scalar quantities, which provide information about a single fault, or they can be a vector containing information about multiple faults. The residuals are then processed in the controller units, making the final decision regarding the fault's presence, type, and location. The decision-making commonly uses thresholds or other statistical approaches to differentiate between fault types. Model-based FDTs are gaining more attention because of their independence from dedicated sensors. However, developing more accurate models of complex electronic systems could sometimes be challenging. Developing an accurate mathematical model is key to accurate FDTs in model-based FDT mechanisms. Inaccurate, highly approximate mathematical models cause false alarms due to inconsistency in the signals generated by the actual system under normal operating conditions and the signals generated by the system models. Alternatively, increasing the accuracy and complexity of the model by adding more details requires substantial computational resources from the main central processor, which is already burdened with the standard control loop calculations. A generalized model-based FDT system is presented in **Figure 6**.

Model-based FDT mechanisms can be qualitative or quantitative. The qualitative approach is constructed as a model in which the input-output relationship of the system is defined as a qualitative function based on different parts of that system. A quantitative FDT approach uses a model in which the input-output relationship of the system is developed as a mathematical function. Qualitative modeling is easy to achieve as these models are easy to develop. However, the quantitative model-based FDT approach is more accurate and robust. Quantitative model-based FDT technique can further be classified into analytical redundancy, parity space, Kalman filters, parameter estimation, and diagnostic observers.

1. Analytical redundancy: This approach compares the output of a system's mathematical model with the corresponding signals obtained from the actual physical system. The difference between the output produced by the modeled system and the corresponding dynamic signals extracted from the physical system is calculated and is called the residual. The main controller can then process these residuals to extract the needed information on the system's health and flag an alarm if the difference between the measured and the predicted value crosses the threshold.

2. Parity space: This approach checks the parity or consistency of the input-output measurements of the system. The residual generated by the parity space is small

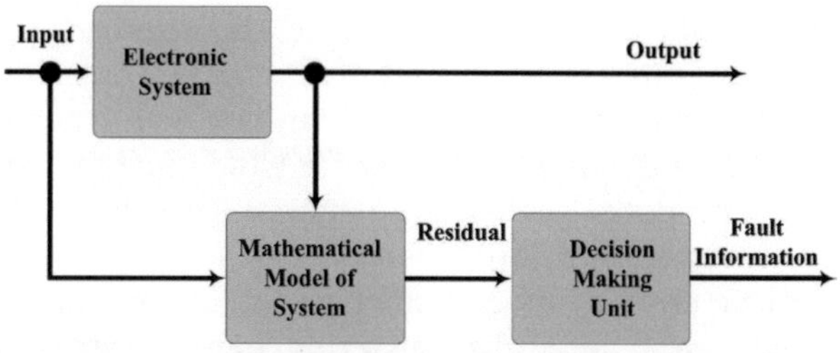

Figure 6.
Model-based FDT technique.

enough to be negligible. However, they become significant under fault conditions, modeling errors, measurement noise, and process noise. An FDT procedure can effectively use the residual from the parity equations to identify and isolate the fault in an electronic system. One specific fault type that the parity space approach can detect is the parametric fault, which is usually associated with semiconductor or AEC degradation.

3. Kalman filters: Calculating the dynamics for measuring the slow deterioration of components in an engineering system is essential for effective fault prognosis. Kalman filter and its variants can effectively estimate the state of the dynamic system parameters under noisy conditions. For example, AEC deterioration can be modeled by a dynamic state-space equation as a dynamic process with time-varying parameters. Next, the slow dynamics of the model are calculated. Kalman filters and their variants can effectively estimate the degraded parameters by handling the slow deterioration process. A threshold is set for the end-of-life parameter, representing the degraded parameters' failure point. The remaining useful life is measured by computing the difference between the failure point and the point of measurement. Similarly, these filters can also be used to estimate the slow degradation of the ON resistance parameter in the semiconductor power switch as an effective fault prognosis tool.

4. Parameter estimation: Parameter estimation is a model-based FDT technique based on the concept that component degradation affects the overall system output and changes the system parameters. A linear parameter-varying model of the system is developed to detect the fault. An interval predictor is applied to the system model, which generates the interval residual. This residual is further utilized to perform FD and FI in the system. To deal with system noise, which may result in a false alarm, a threshold level is computed, and if the residual exceeds this threshold, a fault alarm is triggered, and the accompanying FI is performed.

5. Diagnostic observers: This approach uses different observer types such as Eigen structure assignment observer, Luenberger observer, and fault detection filters for residual generation. The central controller uses this residual for the final FD and FI. Observer-based FDT combines a measured feedback signal with knowledge of the system components. This step helps predict the systems' behavior more precisely than using the feedback signal alone and provides a more accurate estimation of the system behavior under normal and fault conditions.

3.4 Historical data-based fault diagnostics

Generally termed a data-driven FDT technique, this approach develops a model from the known and measured input/output process data. Historical data-based FDT techniques can be a potent mechanism for fault detection and isolation for highly complex non-linear dynamical systems for which accurate mathematical models are difficult to produce. Developing a mapping between an input and output space and creating an associated memory can generate a legitimate output when presented with an unseen input. History-based FDT methods can be categorized into fuzzy logic, neural networks, clustering, expert systems, and pattern recognition. Recently, history-based FDT has gained much interest because of its capability to model complex and non-linear time-invariant systems without worrying about the intricate details of the system for creating accurate models. The data-driven FDT techniques are becoming more easily realizable and more popular as powerful and low-cost GPUs are introduced, which can process these models more accurately.

1. Fuzzy logic: Fuzzy logic can be instrumental when the system cannot be adequately modeled or expressed mathematically. A fuzzy model can be developed based on the knowledge of system behavior or experimental information. This technique is useful for detecting intermittent faults that are difficult to detect using standard FDT schemes.

2. Neural networks: A neural network is an effective historical data-based FDT technique that processes the inputs and outputs of the system to develop a dynamic system model. For example, the model is further trained using first-order gradient algorithms, which represent the original system. The residual sequence can be extracted by calculating the difference between the output measurements of an actual system and the estimated output calculated by the neural network model. This further undergoes statistical processing to develop a stochastic variable that follows the χ^2 distribution. Next, a fault-indicating threshold is calculated, which defines the upper and lower bound of a confidence interval of the χ^2 distribution. The system is considered fault-free if the calculated stochastic variable is within this bound. However, a fault alarm is triggered if this variable crosses the bound. The exact location and type of fault can be identified by repeatedly applying the corresponding statistical test to sub-spaces of the system state-space model.

3. Clustering: Clustering for fault diagnostics in electronic systems involves grouping data points into clusters based on similarity. This unsupervised learning method can help identify patterns or anomalies in the data that correspond to normal or faulty states of the system, even when explicit labels for these states are unavailable.

4. Expert systems: Expert systems is a branch of artificial intelligence that uses a set of predefined rules or a knowledge base to infer and make decisions similar to a human expert. In the case of FDT for electronic systems, it leverages the knowledge base of experienced engineers and domain experts to detect and identify faults.

5. Pattern recognition: Pattern recognition identifies irregularities and anomalies of the FDT signatures. These anomalies can then be utilized to detect the presence of the fault. Each anomaly or irregularity can be associated with a specific fault type and location, which can help achieve the system's final FI.

4. Guidelines for selection of FDT mechanism

This section gives the reader a guideline for selecting an FDT mechanism depending on the conditions and factors. Each mechanism has its strengths and is suited to different scenarios. Factors such as system complexity, desired detection speed, fault type and its criticality, and data availability help the reliability engineer choose the best FDT mechanism.

Hardware-based FDT techniques have the advantage of being reliable and fast for detecting and isolating predefined fault types. They can be used where response time to catastrophic faults is critical, and a lapse in detection and isolation can cause the overall system to shut down. They are implemented using real-time hardware circuits and sensors, which make them suitable for limited predefined fault types. However, this approach also has the downside of limited flexibility in updating the present FDT mechanism to detect new kinds of faults. Moreover, the addition of sensors and corresponding auxiliary circuitry for fault signature detection and measurement may cause the overall cost and size to increase.

Model-based FDT techniques can detect a wide range of faults, including those that are difficult to detect using traditional hardware-based FDT techniques. It can be used when the original system's detailed and accurate mathematical model can be developed. This way, the system's behavior under normal and faulty conditions can be accurately predicted. This approach is effective for incipient or slow drift fault detection where the slow degradation of different parameters of components, such as the equivalent series resistance of the AEC and the ON resistance of the power switch, can cause system malfunction. The hardware-based FDT mechanism cannot accurately detect the subtle changes in these parameters. This approach is more flexible as it could be easily modified to include the detection of new types of faults. However, this approach requires a deep understanding of the system, and sometimes, quantitative modeling of the system could become very difficult and require higher computational resources. Additionally, the slow detection speed of model-based FDTs prohibits its usage for detecting critical, catastrophic faults where detection speed is of the highest priority.

AI-based FDTs are highly flexible and can detect and isolate faults from various categories. They can be used when accurate system modeling is challenging to achieve. This approach effectively detects slow drift faults, which are introduced in the system due to changes in the individual component parameters over a long period. Therefore, these techniques could be used to detect incipient or parametric faults. With the availability of large amounts of historical data under normal operating and fault conditions, the accuracy of the techniques could be further improved by training the AI model. However, they require substantial data for training, which may not be available for all systems, especially under fault conditions. However, this problem can be tackled by using the data generated by the digital twins. The black box approach of the history-based FDT mechanism can make it challenging to understand the basis for a given diagnosis mechanism and may also require significant computational resources, especially for training. Like the model-based FDTs, the slow detection speed of history-based FDTs prohibits its usage in detecting catastrophic faults where detection speed is the highest priority.

5. Conclusions

FT electronic systems are preferred because they continue to operate under fault conditions, possibly at a reduced performance rather than a complete shutdown. This behavior is crucial in critical engineering systems where reliability and continuous operation are paramount. With the integration of effective FDTs, faults can be detected swiftly and accurately. An FT electronic system, bolstered by an effective FDT mechanism, offers enhanced resilience, robustness, and safety. Regarding the selection of the FDT technique, there is no single technique that can serve all the situations. The selection, be it hardware-based, model-based, or data-driven, depends upon the specific application, the ability to model the system accurately, detection speed, the availability of data, and cost considerations. A hybrid approach combining hardware-based, model-based, and data-driven techniques may offer the most robust and comprehensive solutions for effective FDTs across various applications.

Abbreviations

FDT	fault diagnostics
FT	fault tolerant/fault tolerance
FD	fault detection
FI	fault isolation
AEC	aluminum electrolytic capacitors
SCF	short circuit faults
OCF	open circuit faults
TDDB	time dependent dielectric breakdown
NBTI	negative bias temperature instability
HCI	hot carrier injection
FDI	fault detection and isolation
EMI	electromagnetic interference

Author details

Shahamat Shahzad Khan[1,2]

1 University of Liverpool, Liverpool, UK

2 Xian Jiaotong Liverpool University, Suzhou, China

*Address all correspondence to: shahamatkhan@yahoo.com

References

[1] Gray J, Siewiorek DP. High-availability computer systems. Computer. 1991;**24**(9):39-48. DOI: 10.1109/2.84898 [Accessed: December 18, 2024]

[2] Yang S, Bryant A, Mawby P, Xiang D, Ran L, Tavner P. An industry-based survey of reliability in power electronic converters. IEEE Transactions on Industry Applications. 2011;**47**(3):1441-1451. DOI: 10.1109/TIA.2011.2124436 [Accessed: December 18, 2024]

[3] wikipedia.org. Fault Detection and Isolation. San Francisco, California: Wikipedia; 2023. [Online]. Available from: https://en.wikipedia.org/wiki/Fault_detection_and_isolation [Accessed: December 22, 2024]

[4] Nexperia. Failure signature of electrical overstress on power MOSFETs- Application Note:11243. 2017. Available from: https://assets.nexperia.com/documents/application-note/AN11243.pdf [Accessed: December 19, 2024]

[5] Pandya K, Pietrczak K, Chiang A, Getzan G. Power MOSFET Failures in Automotive Application- VISHAY SILICONIX, Application Note 910. Singapore; 2009. Available from: https://api.semanticscholar.org/CorpusID:12163743 [Accessed: December 19, 2024]

[6] Vernica I, Ma K, Blaabjerg F. Optimal derating strategy of power electronics converter for maximum wind energy production with lifetime information of power devices. IEEE Journal of Emerging and Selected Topics in Power Electronics. 2008;**6**(1):267-276. DOI: 10.1109/JESTPE.2017.2745714 [Accessed: December 18, 2024]

[7] Tripathi H, Pradhan N. Reliability optimization of electronics module by derating using genetic algorithm. In: 2016 International Conference on Control, Computing, Communication and Materials (ICCCCM); 2016; Allahbad, India. Piscataway, NJ, USA: IEEE Power Electronics Society; 2016. pp. 1-6. DOI: 10.1109/ICCCCM.2016.7918264 [Accessed: December 18, 2024]

[8] Pei X, Nie S, Chen Y, Kang Y. Open-circuit fault diagnosis and fault-tolerant strategies for full-bridge DC–DC converters. IEEE Transactions on Power Electronics. 2012;**27**(5):2550-2565. DOI: 10.1109/TPEL.2011.2173589 [Accessed: December 18, 2024]

[9] Khan S, Wen H. A comprehensive review of fault diagnosis and tolerant control in DC-DC converters for DC microgrids. IEEE Access. 2021;**9**:80100-80127. DOI: 10.1109/ACCESS.2021.3083721 [Accessed: December 18, 2024]

[10] Jamshidpour E, Poure P, Saadate S. Photovoltaic systems reliability improvement by real-time FPGA-based switch failure diagnosis and fault-tolerant DC–DC converter. IEEE Transactions on Industrial Electronics. 2015;**62**(11):7247-7255. DOI: 10.1109/TIE.2015.2421880 [Accessed: December 18, 2024]

[11] Siouane S, Jovanović S, Poure P. Open-switch fault-tolerant operation of a two-stage Buck/Buck–boost converter with redundant synchronous switch for PV systems. IEEE Transactions on Industrial Electronics. 2019;**66**(5):3938-3947. DOI: 10.1109/TIE.2018.2847653 [Accessed: December 18, 2024]

[12] Cordeiro A, Pires VF, Silva JF. Fault-tolerant design of three-phase dual-buck VSI topologies. In: 45th Annual Conference of the IEEE Industrial Electronics Society (IECON 2019); 2019; Lisbon, Portugal. Vol. 1(1). Piscataway, NJ, US: IEEE Industrial Electronics Society (IES); 2019. pp. 4630-4635. DOI: 10.1109/IECON.2019.8927183 [Accessed: December 18, 2024]

[13] Benyamina F, Khoucha F, Benrabah A, Djenadbia A, Benbouzid M. Experimental evaluation of power switch fault-tolerant control of a six phase mirror interleaved boost converter for fuel cell applications. In: 2019 International Conference on Advanced Electrical Engineering (ICAEE); 2019; Algiers, Algeria. Vol. 1(1). Piscataway, New Jersey, United States: IEEE Society; 2019. pp. 1-6. DOI: 10.1109/ICAEE47123.2019.9014744 [Accessed: December 18, 2024]

[14] Haghbin S, Blaabjerg F, Bahman AS. Frozen leg operation of a three-phase dual active bridge converter. IEEE Transactions on Power Electronics. 2019;**34**(5):4239-4248. DOI: 10.1109/TPEL.2018.2865092 [Accessed: December 18, 2024]

[15] Wang Y, Chen S-Z, Wang Y, Zhu L, Guan Y, Zhang G, et al. A multiple modular isolated DC/DC converter with bidirectional fault handling and efficient energy conversion for DC distribution network. IEEE Transactions on Power Electronics. 2020;**35**(11):11502-11517. DOI: 10.1109/TPEL.2020.2985232 [Accessed: December 18, 2024]

[16] Song W, Huang AQ. Fault-tolerant design and control strategy for cascaded H-bridge multilevel converter-based STATCOM. IEEE Transactions on Industrial Electronics. 2010;**57**(8):2700-2708. DOI: 10.1109/TIE.2009.2036019 [Accessed: December 18, 2024]

[17] Tu P, Yang S, Wang P. Reliability- and cost-based redundancy design for modular multilevel converter. IEEE Transactions on Industrial Electronics. 2019;**66**(3):2333-2342. DOI: 10.1109/TIE.2018.2793263 [Accessed: December 18, 2024]

[18] Liu J, Zhao N. Improved fault-tolerant method and control strategy based on reverse charging for the power electronic traction transformer. IEEE Transactions on Industrial Electronics. 2018;**65**(3):2672-2682. DOI: 10.1109/TIE.2017.2748032 [Accessed: December 18, 2024]

[19] Zhang Y, Yang R, Wen H. JITL based fault detection of DAB DC-DC converter with EPS control. In: 2020 39th Chinese Control Conference (CCC); 2020; Shenyang. China. 2020. pp. 6208-6213. DOI: 10.23919/CCC50068.2020.9188659 [Accessed: December 18, 2024]

[20] Zhuang Y, Liu F, Zhang X, Huang Y, Zha X, Zhouyang L. Short-circuit fault-tolerant topology for multiport cascaded DC/DC converter in photovoltaic power generation system. IEEE Transactions on Power Electronics. 2021;**36**(1):549-561. DOI: 10.1109/TPEL.2020.3004070 [Accessed: December 18, 2024]

[21] Soon JL, Lu DD-C, Peng JC-H, Xiao W. Reconfigurable nonisolated DC–DC converter with fault-tolerant capability. IEEE Transactions on Power Electronics. 2020;**35**(9):8934-8943. DOI: 10.1109/TPEL.2020.2971837 [Accessed: December 19, 2024]

[22] Jiang X, Li Q, Huang W, Cao R. A dual-winding fault-tolerant motor drive system based on the redundancy bridge arm. IEEE Transactions on Industrial Electronics. 2019;**66**(1):654-662. DOI: 10.1109/TIE.2018.2833023 [Accessed: December 19, 2024]

Chapter 5

The System of Spatial Recognition in the Brain

Keiko Tsujioka

Abstract

Recent discussions have focused on flight training as a preventive measure against accidents caused by spatial disorientation (SD). This is because existing simulation systems for mechanical equipment rely on experienced-based training rather than fostering self-generated attitude change; that is, they emphasize a hands-on educational approach. This might be a real problem for pilots. Accordingly, in this chapter we study a simulation system using three-dimensional computer graphics that can help novice pilots act from shared awareness with an AI robot by analyzing themselves from self-generated attitude change. This AI robot, known as HARIS (Human-AI Interactive Support System), interactively supports novice pilots within the simulation system, adapting its guidance to each pilot's personality and cognitive traits.

Keywords: spatial disorientation (SD), human-AI robot interactive support system (HARIS), stress tolerance of IIV-RT (intra-individual variability in reaction time), dual-task, cognitive traits, personality

1. Introduction

Spatial orientation refers to how humans recognize their own location, position, and direction. It is a fundamental element of human behavior. When this state is impaired or difficult to maintain, it is typically regarded as spatial disorientation (SD) [1]. In an aeronautical environment, SD occurs when pilots are unaware of the correct position, movement, or attitude of their aircraft.

Because the emphasis is on educating the preconscious awareness of SD and its separation from the physical plane, the best course of action is to operate under instrument flight rules (IFR). About 80% of catastrophic accidents are caused by unawareness of SD, which is defined as Type 1 in the three types of psychological types. Other types of SD are caused by various illusions across a wide range of biological taxonomy, such as optic or labyrinthine ataxia [2]. Therefore, various reproducible training devices and alarm systems have been developed in simulator systems.

Investigations into airplane accidents have revealed links to stress as a human factor. Despite ongoing training efforts, SD-related accidents continue to occur, raising questions about the effectiveness of current preventive measures [3]. This may be due to the lack of objective data analysis supporting their efficacy [4]. Previous studies have relied on subjective questionnaires administered to pilots after accidents to explore the relationship between SD and incident occurrence.

To clarify this issue, the following questions arise:

- Why do pilots experience SD?
- What conditions are related to accidents?
- Why has training effectiveness not been shown?
- What accounts for differences in training effectiveness between experienced and novice pilots, with the former performing better? [5]
- How have preventive measures, such as training for self-regulatory skills, progressed?

From these investigations, new insights have emerged:

- While research on SD has focused on external factors, internal aspects, like psychological factors and individual traits, remain unclear in terms of their relationship to SD.
- Although SD-related accidents caused by unawareness or stress are cited, there are few reports addressing stressor scales, stress tolerance measurement techniques, or stress control. However, some studies have examined stressor discrimination among pilots compared to other professions [6, 7].
- Regarding differences between experienced and novice pilots [8], the third layer of awareness proposed by Osaka may be relevant [9]. The first layer is considered an activator, akin to booting up a computer, and related to attention and perception. The second layer facilitates detection of illusions through parallel processing, helping experienced pilots remain situationally aware. On the other hand, novice pilots may be more prone to SD due to limitations in the third layer's simultaneous processing functions, which may hinder situational awareness [10, 11].

To address these problems, it is suggested that training should incorporate individual characteristics and cognitive traits. Based on this reasoning, we propose the following research questions:

- Are there relationships between individual differences in traits, such as variations in language information processing [12–14], and stress?
- How can pilots control their own stress consistently?
- Are there methods for measuring stress tolerance that allow pilots to manage themselves using predictable patterns [15, 16]?
- What are the reasons for stress control using objective indicators?

From those viewpoints, we will develop a measuring instrument for stress tolerance based on intra-individual variability in reaction time (IIV-RT). This tool may help pilots become aware of their own stress levels and improve their stress management, control, and coping strategies by providing individualized measurement results [17]. It is expected that pilots will be able to autonomously control stress, enabling early awareness of SD and thorough checks via instrument flight [2].

Given that stress may impair working memory and contribute to SD, pilots must manage and regulate their stress effectively. Therefore, it is essential for both pilots and the AI support system to analyze stress levels consistently. Stress tolerance levels [18–21] are predicted to fluctuate depending on the nature and intensity of stressors. For this reason, it is vitally important to measure stress tolerance before each duty. This may enhance pilots' self-regulatory skills, such as conscious stress coping and management [22]. In addition, this stress tolerance measurement system will be developed as part of HARIS (Human-AI Interactive Support System), which aims to support pilot training for the prevention and self-management of SD.

In this study, we analyzed response times from the presentation of psychological questionnaires on personality [23] to the time when participants selected their answers, in order to clarify aspects of human information processing [24]. These data revealed individual differences in cognitive traits. During the decision-making process, emotion and memory were reentered through feedback and feedforward control. These phenomena showed clear cognitive traits depending on type.

Separately, studies on the causes of stress have clarified the relationship between the hippocampus and emotion [25–27]. These findings suggest that stress affects memory, particularly working memory.

From this viewpoint, IIV-RT is likely influenced by various factors related to information processing [28]. Therefore, we propose a method for measuring stress tolerance using IIV-RT through a dual-task approach [29], in addition to traditional psychological questionnaires. This method imposes a greater cognitive load, inducing stronger stress than traditional approaches. It is predicted that this will affect reaction time [12–14].

For this reason, analysis of stress tolerance using IIV-RT should be based on previous study methods, enabling identification of individual cognitive traits as criteria

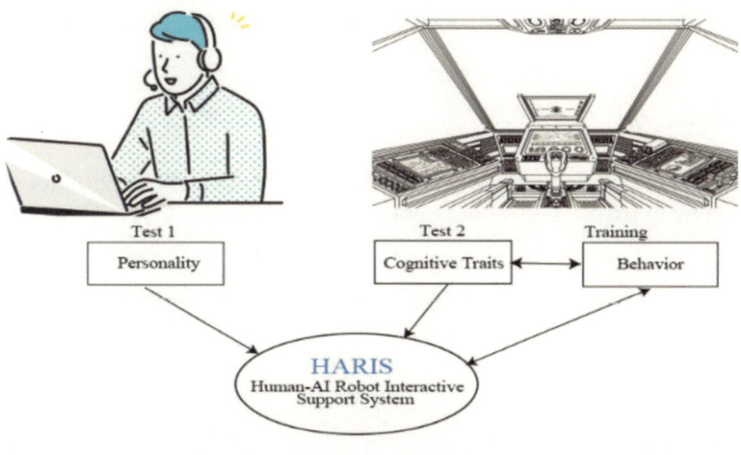

Figure 1.
Conceptual diagram of HARIS (Human-AI Robot Interaction Support System).

and standard values [12, 14, 30]. The procedure will involve comparing results to determine whether other influencing factors are present. Through these objective measurement systems, it may be possible to improve countermeasure training and prevent SD-related accidents (**Figure 1**) [4, 31–33].

2. Instruction sheet

2.1 Aims

In this study, training for accident prevention from SD will be reexamined, and a new method will be proposed. The new method, HARIS, will be developed and implemented.

2.2 Objective

The objective of this study, through the development of HARIS, is to prevent accidents by investigating the unresolved causes of SD.

2.3 Strategy

It is considered that the cause of SD associated with psychological factors is unawareness, which is strongly connected to accidents under stress [34, 35]. As a strategy, psychological testing will be applied to investigate how stress may affect pilots.

First, as a criterion for stress tolerance, individual traits of pilots will be measured (**Figure 2**, Test 1). Next, each participant will undergo testing to measure stress tolerance through intra-individual variability in reaction time (IIV-RT) (**Figure 2**, Test 2). Stress tolerance levels indicate how strongly stress is felt by an individual and how well they may be able to control it.

This stress tolerance level may vary erratically from day to day depending on personality and cognitive traits. For this reason, it should not be compared across individuals but measured within each individual. Accordingly, multiple sessions will be conducted to measure IIV-RT stress tolerance and perform simulation training experiments. The results of measurement will be reported to pilots before simulation training [36].

2.4 Implementation

The simulation training for prevention of SD accidents will be conducted using HARIS. Data from psychological testing (criterion) (**Figure 3**, Test 1) and stress

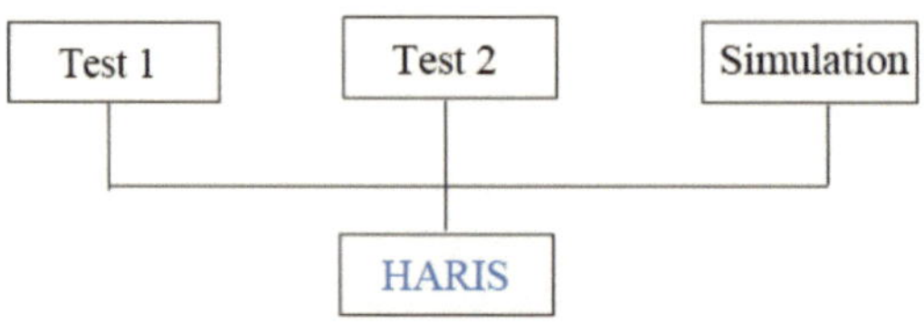

Figure 2.
Architecture diagram of HARIS (Human-AI Robot Interaction Support System).

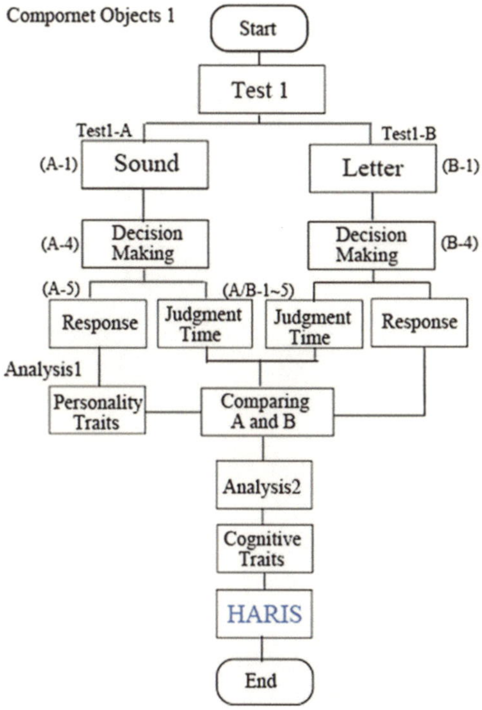

Figure 3.
The first component object used to measure auditory response time in the experiment.

tolerance levels will be sent to HARIS. These data, along with training records and psychological testing results, will be assessed for validation (**Figure 4**, Test 2). As a method of verification, the results of measurements and record data will be analyzed to assess individual stress and awareness objectively. Based on the analysis, the AI will evaluate whether pilots have coped with, managed, or controlled their stress, comparing these evaluations with the results of simulation training.

Specifically, when a pilot faces a dangerous situation, HARIS will help them become aware of it in a way tailored to their individual traits. This situation will also be recorded and evaluated.

2.5 Features of HARIS

Accordingly, the development of HARIS will address these research issues. It is predicted that practical measures for preventing SD-related accidents will advance through objective data assessment by HARIS. HARIS differs from previous training measures by supporting each pilot according to their cognitive traits, drawing on their strengths and helping them fully utilize their abilities. For this reason, collaboration with participants is essential in this research and development, as the stress tolerance measurement is a key component.

Specifically, this research may clarify language information processing more than before by using psychological questionnaires along with response time measurements. It may also advance understanding of human information processing.

On the other hand, Kasuga et al., developed questionnaire methods for technostress tolerance. They analyzed their success using factor analysis of

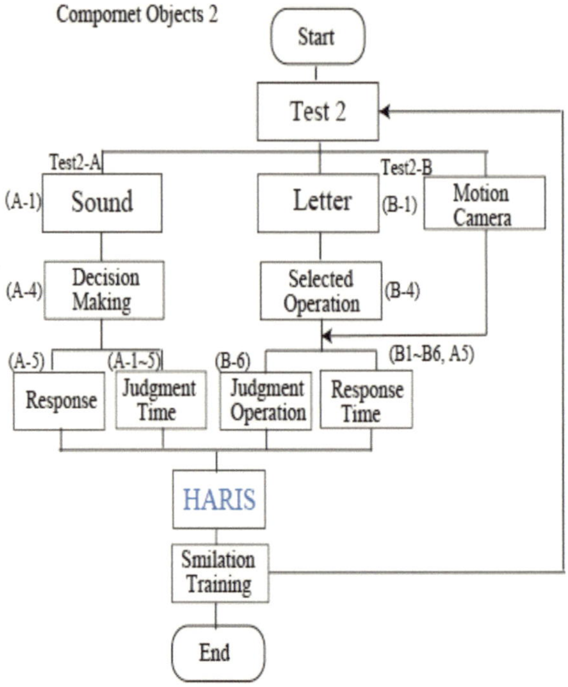

Figure 4.
The second component object used to measure auditory response time in the experiment.

yatabe-guilford personality inventory (YGPI) and concluded that their measurement has high validity. From this viewpoint, our method might also be able to measure the degree of stress tolerance effectively.

Moreover, to measure reaction time objectively through participants' movements, depth cameras will be installed. Because our measurement method requires dual tasks to determine stress tolerance, it is designed to include a testing method known as visual masking [22].

In visual masking, for example, a masked word prevents subjects from becoming aware of it, depending on how the interfering stimulus is presented. The purpose of this experimental method is to investigate differences in perception between conscious and unconsciousness processing.

This method may be comprehensible to participants, but if they adopt a serious stance, the data obtained should be highly accurate. In addition, if participants provide feedback or suggestions for improvement, the system can be modified accordingly. It is predicted that participants' condition changes, depending on their physical and psychological state, will be observable through this measurement method. We aim to investigate the connection between these changing conditions and simulation performance.

As part of the development and validation procedure, participants are required to respond to psychological testing questionnaires presented via sound or text using a traditional method (**Figure 3**). Next, they will undergo stress tolerance measurement (**Figure 4**), followed by a simulation exercise (**Figure 1**).

At every step, participants will complete additional questionnaires before and after both the measurement and simulation training. Test 1 (YGPI) will be conducted only during the first session because its purpose is to gather baseline data for criteria analysis. From the second session onward, participants will be examined using Test 2 and simulation training (**Figure 2**). Throughout the experiment, participants must follow instructions delivered via voice and respond accordingly by answering or navigating.

For the psychological testing, participants select responses of "Yes," "No," or "?" based on comparisons with their everyday experiences. Detailed explanations of each test will be provided at every session. Performance results will be presented graphically (**Figure 4**). During each training session, participants are expected to act based on insights gained from these results. It is expected that their stress coping, management, and control will improve through continued engagement. As a result, their stress tolerance levels should increase, contributing to accident prevention.

2.6 Test 1: Dual tasks 1

Objective: Measure personality and cognitive traits.

Goal: Establish individual criteria values for stress tolerance levels.

Testing contents: Measure both response time and selected answers using questionnaires presented via sound or text.

Method:

- 120 psychological testing items are presented by sound (**Figure 3**, Test 1 A-1) or by text (**Figure 3**, Test 1 B-1)
- Participants selected answers and response times are recorded
- Personality traits are assessed based on selected answers
- Cognitive traits are determined by comparing response times across both modalities (**Figure 3**)

2.7 Test 2: Dual tasks 2

Objective: Strengthen participants' stress coping, management, and control.

Goal: Enhance conscious awareness during SD.

Testing contents: Measure stress tolerance through IIV-RT.

Method:

- Reaction time is measured through a dual-task approach involving a primary and secondary task
- Participants monitor their own stress tolerance variability daily to improve coping, management and control
- This process is expected to help participants recognize when they are experiencing SD

Measuring method:

Primary task (**Figure 4**, Test 2-A):

- The same 120 psychological testing items are presented (**Figure 4**)
- Participants respond by selecting among "Yes," "No," or "?" (**Figure 4**, Test 2 B-6)

- Response time is measured from the onset of the voice prompt to the completion of the response (**Figure 4**, Test 2 B-6)

Secondary task (**Figure 4**, Test 2-B):

- Answer selection methods are displayed on the Heads-Up Display (HUD)
- Participants must choose the appropriate operation based on their selected answer from the voice prompt (**Figure 4**, B-6)
- Participant motion is measured for each questionnaire item
- All data will be transmitted to HARIS for analysis (**Figures 1** and **2**)

2.8 HARIS

The data gathered will be sent to HARIS. The analysis results will be displayed on screens, and the AI (Mr. HARIS, **Figure 4**) will explain to each participant their individual degree of stress tolerance and IIV-RT.

Next, participants will engage in simulation training based on the findings from their own results. During training, if a participant begins to show patterns associated with SD, HARIS will issue a caution and give hints via sound prompts, such as guidance related to IFR. If participants recognize these hints and change their behavior accordingly, the effectiveness of their stress control, coping, and management will be evaluated.

3. Method

3.1 Outline of data gathering and method

3.1.1 Rationale for developing measurement methods: Stress tolerance level in IIV-RT

Research on stress was pioneered by Selye [33], followed by Cannon [34], who discovered that the biological body exhibits reflexive responses to various stimuli. These foundational studies contributed significantly to the medical understanding of the relationship between stress stimuli and physiological responses. At a certain level, stress has also been considered beneficial for learning, particularly due to the role of adrenaline.

In the 1990s, psychological testing gained momentum, especially in evaluating pilot traits. This increase may have been driven by the emergence of new types of aircraft along with rapid advances in computer systems, prompting the need to assess pilot aptitude. Aptitude testing was developed to evaluate individual traits and determine whether pilots could maintain situational awareness and make decisions under changing conditions. This led to increased interest in the relationship between awareness, capacity of working memory, and stress.

At the same time, technological progress spurred investigations into workplace-related stress, particularly in the context of mental health. Researchers began to recognize that mental health must be assessed differently depending on workplace environments because stress models vary significantly. This shift introduced the concept of technostress and efforts to develop methods for measuring technostress tolerance.

Pilots, in particular, are exposed to intense and specific stressors. When unexpected events occur, they are required to respond swiftly to emergency situations. For novice pilots, environmental factors, especially the cockpit setting, have been identified as highly stressful, with human factors often cited as the primary cause.

In such cases, managing severe stress becomes crucial, especially as it affects pilots' awareness of SD. In a study by Kasuga et al., researchers developed questionnaires related to technostress syndrome and measured the degree of stress tolerance. They reported that participants improved their stress control, management, and coping by reflecting on their measurement results, which proved to be a successful approach.

From these viewpoints, our study provides each pilot with a report on their individual degree of stress tolerance prior to undergoing SD measurement simulation in a cockpit environment. This feedback is expected to enhance their ability to manage, control, and cope with stress. In addition, we propose a framework for how HARIS supports this process, including the methods used for measurement and analysis.

3.1.2 Method for measuring stress tolerance criteria

3.1.2.1 Method of operation: Test 1

1. Initial instruction

   Participants will first receive a comprehensive explanation of the entire procedure.

2. Testing guidelines

   Participants will be informed of key points to keep in mind during testing. Although the same set of 120 questions will be presented, either by audio or text, responses may vary. Participants should concentrate on answering each current question (**Figure 5**).

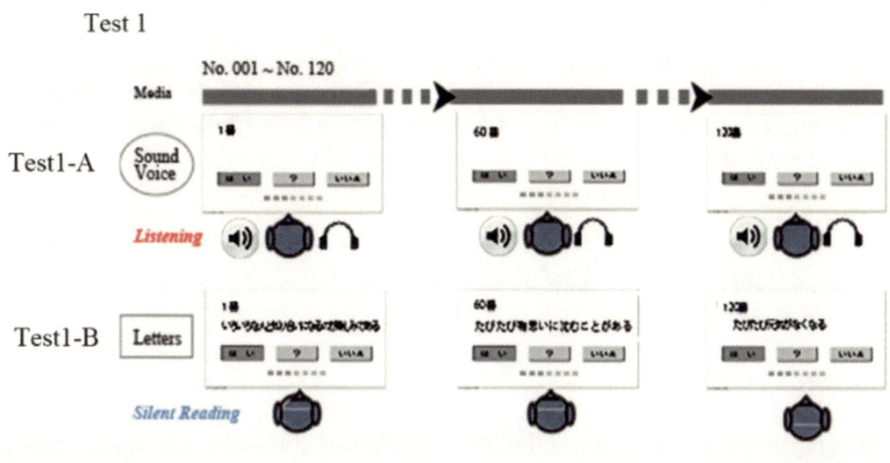

Figure 5.
Testing method for scenarios 1-A and 1-B.

3. Response instructions

   For each question, participants should respond based on whether the content reflects their daily behavior:

   If it matches, select "Yes"

   If it differs, select "No"

   If uncertain, select "?"

   Participants are advised not to overthink their responses. Whenever possible, they should choose either "Yes" or "No" to avoid indecision.

4. Answer selection

   Responses are entered via keyboard using the following keys:

   1 = "Yes"

   2 = "No"

   3 = "?"

5. Visual feedback

   Once a button is selected, it will be visually highlighted to confirm the choice.

6. Time restriction and progress indicator

   Participants must respond within a set time limit. A progress bar will indicate elapsed time and signal when the screen will transition to the next question.

7. Screen transition time

   Each question screen will fade in with the question number. After 3 seconds of audio presentation, the screen will fade out. This fade-in/fade-out mechanism is designed to reduce visual stress. The same procedure applies to text-based presentations.

8. Testing sequence

   The test will begin with audio-based questions, followed by the text-based version.

9. Missed responses

   If a participant fails to respond within the allotted time, the same question will be repeated immediately after the previous one.

10. Post-test survey

    Upon completing both tests, participants will be asked to fill out a brief survey regarding their daily habits. For example, "How many hours did you sleep today?"

3.1.2.2 Method of data gathering: Test 1

After participants are informed about the purpose of testing, the testing method, the reporting of results, and the support system, we will obtain informed consent. Participants will then be asked to enter their personal information to confirm their identity on the screen (including ID and password).

"Response time" is defined as the time measured from the start of questionnaire presentation to the moment the response button is highlighted (by operating the keyboard). "Response" refers to the button that has been highlighted and selected by the participant. Each participant completes psychological testing consisting of 120 questions presented via sound and 120 via text, resulting in a total of 240 (**Figure 5**) [37] responses and corresponding response time data per person.

In addition to this data, surveys will be conducted at the pre-, post-, and intermediate stages of testing. These survey responses will also be collected for analysis.

3.1.2.3 Method of data analysis: Test 1

A prototype experiment involving 28 participants was conducted using the same measurement method described in the previous section [37]. Following this, a practical experiment with 98 participants was carried out. In this section, we present the method of data analysis, along with the criteria used to determine degrees of stress tolerance based on the experimental results.

Under consistent conditions and quality standards, measurements of responses and response times for cognitive traits were conducted twice, with a 3-month interval between sessions. The resulting data and analysis verified the reliability of reexamining this measurement method.

Subsequently, practical experiments were conducted under the same conditions prior to collaborative learning. Based on the measurement results, reliability and validity were assessed by comparing effectiveness among matched participants, optimizing the collaborative leaning environment according to cognitive traits. Details of the analysis method and corresponding results are presented below.

3.2 Prototype experiment

3.2.1 Validation of response time measurement

To validate the response time measurement, two experiments were conducted 3 months apart under identical conditions and quality standards. In each session, participants responded to questionnaires presented either as auditory stimuli (spoken voice) or visual stimuli (written letters).

All questionnaire items consisted of single short sentences. These were first categorized into groups based on the number of words per sentence. Response times were then compared among groups, with separate analyses for each presentation

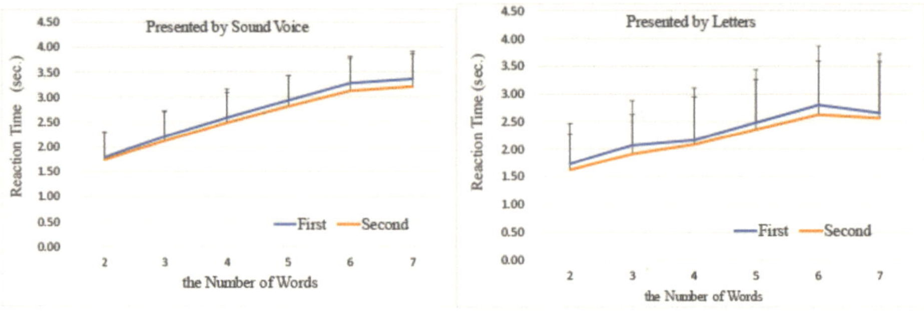

Figure 6.
Response time to 120 questionnaires divided by the number of words each, presented by sound voice (left) and by letters (right), comparing between the first and second times.

(**Figure 6**). The resulting graphs indicate no significant differences in response times between the first and second experiments, regardless of whether the stimuli were presented as sound or text (**Tables 1** and **2**). These results confirm the reliability and reproducibility of the measurement system, showing a high level of calibration suitable for scientific application.

3.2.2 Analysis of response time (language information processing)

3.2.2.1 Validation of analysis method 1 (cognitive style)

As shown in the previous section's graphs, the correlation coefficients between response time and the number of words did not reveal significant differences when comparing sound voice and letter presentations across the first and second measurements.

However, differences were observed in the correlation coefficients for response time between sound voice (r = 70) and letters (r = 37) (**Table 3**). In addition, the standard

	Correlation coefficient					
	Materials of media					
	Sound voice			Letters		
	df.	*r*	*t-Ratio*	*df.*	*r*	*t-Ratio*
Pre (N = 28)	3360	0.70	*n.s.*	3360	0.37	*n.s.*
Post (N = 28)	3360	0.68		3360	0.38	

Table 1.
Correlation coefficient between pre-post average response time to questionnaires by presenting sound voice or letters.

Cognitive style	Correlation coefficient
Visual type	r < 0.3
Auditory type	r > 0.5
Intermediate type	0.3 =< r <=0.5

Table 2.
Correlation coefficient of response time between presenting sound voice and letters depending on cognitive styles.

DOI: http://dx.doi.org/10.5772/intechopen.1011966

	Comparison of response time							
	Materials of media							
Traits of information processing I	**Sound voice**				**Letters**			
	$\bar{X}$ (*SD*)	*df.*	*t-Ratio*	*t-Test*	$\bar{X}$ (*SD*)	*df.*	*t-Ratio*	*t-Test*
Auditory type (N = 31)	2.64 *(0.89)*	5156	2.12	*	2.65 *(0.98)*	5154	−21.05	***
Visual type (N = 13)	2.70 *(0.82)*				2.01 *(0.92)*			

*Significant differences (*t < 0.05, **t < 0.01, ***t < 0.001).*

Table 3.
Comparison of average response time to questionnaires by presenting sound voice or letters between auditory and visually processing type.

deviation for letter-based responses was greater than for sound voice. Specifically, as the number of words increased, the standard deviation of response time for letters also increased. This suggests greater individual differences in language information processing when questionnaires are presented in written form compared to auditory form.

Building on this analysis, participants in the practical experiment were grouped based on their correlation coefficients between word count and response time during letter-based presentations (**Table 3**, **Figure 6**). Two cognitive processing types were identified:

Visual type ($r < 0.3$)
Auditory type ($r > 0.5$)

These classifications were used to assess reproducibility, consistent with the prototype experiment (**Table 2** and **Figures** 7 and **8**).

These findings support the validity of the analysis method for categorizing participants based on cognitive style, which can serve as a criterion for assessing degrees of stress tolerance. Additionally, learning effects revealed interaction effects between the two types (**Tables 4** and **5**, **Figure 9**). Although the Auditory type had lower average scores than the Visual Type in the first semester, this trend reversed in the second semester. This suggests temporal variation in performance across cognitive styles.

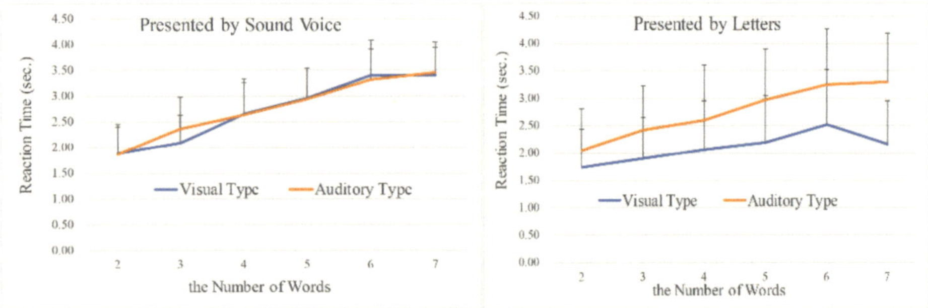

Figure 7.
Comparison of response time to 120 questionnaires presenting by letters or sound voice, grouped by word count between visually and auditory processing types.

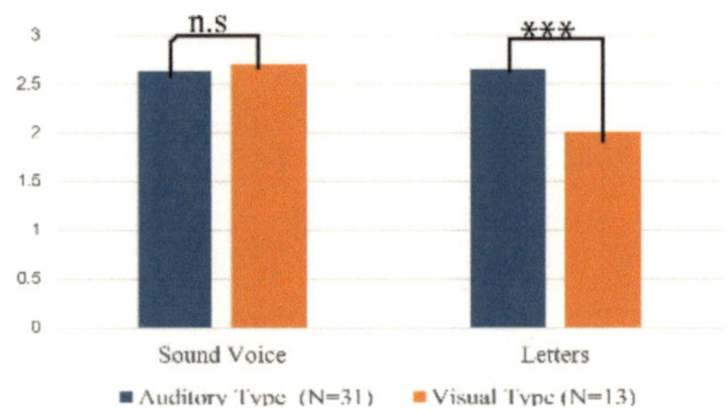

Figure 8.
Comparison of response time to 120 questionnaires presented by letters or sound voice, divided by word count between visual and auditory processing types.

Cognitive type	Number	Average score (100 points)		SD	
		First semester	Second semester	First semester	Second semester
Visual type	12	85.9	86.0	12.6	5.3
Intermediate type	55	83.9	86.7	11.2	7.2
Auditory type	31	83.4	88.4	14.0	3.8

Table 4.
Differences of learning effects among cognitive types by comparing average scores between the first and the second semester.

Element of material	Context	Items	Minimum	Maximum	x	SD
The number of characters	Emotional	60	4	20	14.3	4.2
	Non-emotional	60	4	20	14.2	4.2
The length of time for reading item	Emotional	60	0.76	3.2	2.0	0.6
	Non-emotional	60	0.73	3.0	2.0	0.6

Table 5.
Comparison of the number of characters or the length of time for reading item between emotion /non-emotion in 120 questionnaires.

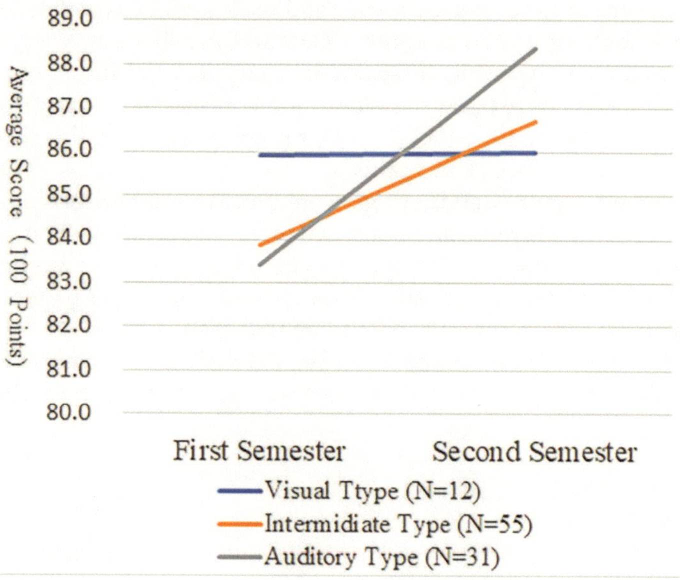

Figure 9.
Differences of learning effects among cognitive types by comparing average scores between the first and the second semesters.

3.2.2.2 Validation of analysis method II (cognitive traits)

Next, the response time data for letter-based questionnaire presentations were analyzed based on the content of the questions and compared across participants. Participants were asked to reflect on their daily behaviors in relation to the questionnaire items. This required them to retrieve episodic memories and compare those experiences with the questionnaire content. Therefore, the responses were based not on academic knowledge or cognitive ability but on personal experience.

It was anticipated that individual differences in information processing would emerge, depending on how each participant accessed and evaluated their memories. From this perspective, 120 questionnaire items were divided into two categories based on content: emotional and non-emotional contexts, with 60 items in each category. Both reading time (via sound presentation) and response time (via letter presentation) were categorized accordingly (**Table 5**).

Then, response times for emotional and non-emotional items were compared with their respective reading times. The correlation coefficients between response time and reading time were calculated for both emotional and non-emotional contexts (**Table 5**). The results revealed differences in average correlation coefficients between the two categories. Moreover, there were differences among participants. Based on these differences, participants were classified into two cognitive trait types: Eidetic Type and Adjusting Type. These classifications were then used to verify distinctions in information processing styles (**Table 6**) [38, 39].

3.2.2.3 Results of verification II (comparison of learning effects)

As shown in **Figure 10** (left), the Eidetic Type exhibited greater dispersion in correlation coefficients between response time and reading time in both emotional and non-emotional contexts. In contrast, the Adjusting Type showed stronger and more consistent correlations, especially in emotional contexts. In addition, response times for emotional contexts were faster than for non-emotional contexts among Adjusting Type participants (**Figure 10**, right) [24]. This phenomenon was also observed among other Adjusting Type participants.

As such, the two cognitive style groups were investigated for learning effects [13, 30]. As shown in **Figure 11**, an interaction effect was found in their performance scores. For the Eidetic Type, the average score in the first semester was higher than in the second. In contrast, the Adjusting Type showed a crossover pattern, with their average score in the second semester exceeding that of the first.

Next, performance differences between Team B and Team C in a collaborative learning class were analyzed. Participants were grouped into teams based on

Cognitive traits	
Eidetic type	$X = < \mu - \sigma$
Adjusting type	$X = > \mu + \sigma$
Intermediate type	$\mu - \sigma < X < \mu + \sigma$

Table 6.
Differences of correlation coefficients between response time and reading time in emotional or non-emotional context among cognitive traits.

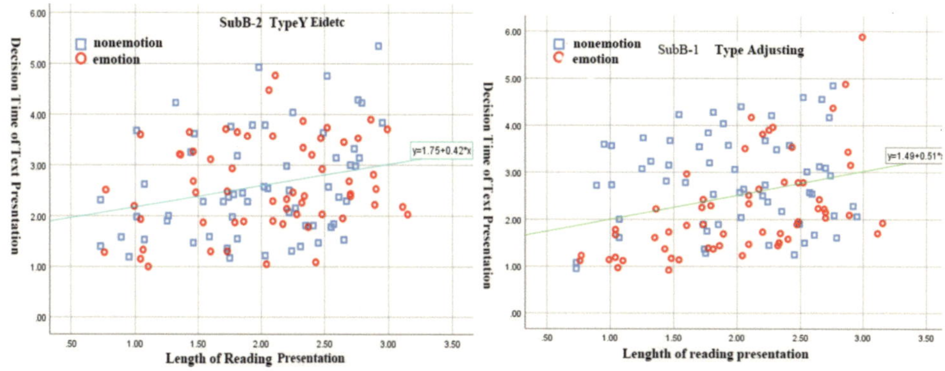

Figure 10.
Comparing response time of Eidetic Type (left) and Adjusting Type (right) between presenting questionnaires by letters, grouped by reading presentation length.

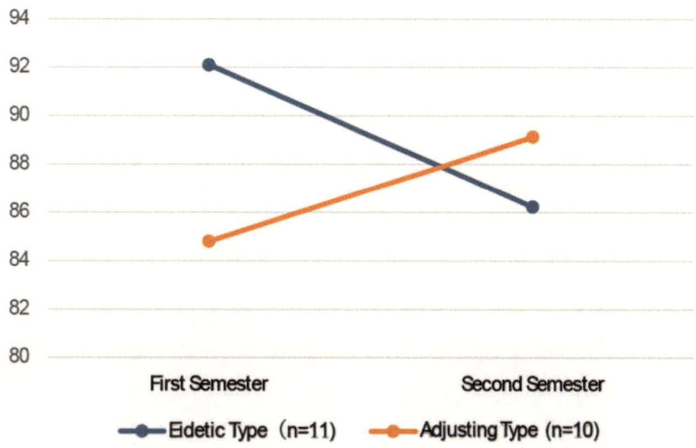

Figure 11.
Comparison of learning effects differences between the first and the second semester, depending on Eidetic Type and Adjusting Type.

optimized matching of individual characteristics. Pre- and post-quizzes were administered during the first semester of the practical class.

The quiz results showed that Team B's average post-quiz scores improved compared to the pre-semester scores. In contrast, Team C's average scores declined relative to their pre-semester performance. However, when comparing final examination scores between the first and second semesters, both SubB-1 and SubC-3 showed significant improvement (**Table 7**).

3.2.2.4 Results of verification II (comparison among personality factors)

Building on the previous section, the graphs show correlations between personality traits and additional factors: lack of Cooperativeness and lack of Objectivity (Co, O; **Figure 12**), as well as Nervousness and Cycle Tendency (N, C; **Figure 13**) for each participant.

For the Adjusting Type participants (SubB-1 and SubC-3), the graphs show low levels across both Co and O factors, as well as low levels of N and C factors. In contrast, the Eidetic Type (SubC-2) showed high levels in both sets of factors.

Participant	Personality type			Pre-post-test			2015 semesters		Differences	Information processing		Correlation coefficient	
	1	2	3	Pre-test	Post-test	Average	First	Second		Types	Traits	Non-emotion	Emotion
SubB-1	C	D	CD	0.5	9.5	9.0	79	97	18	Intermediate	Adjustment	0.09	0.54
SuaB-2	A″	B′	Q				93	91	−2	Visual	Eidetic	0.32	0.22
SubB-3	AB	A	A′				86	91	5	Visual	Intermediate	0.24	0.35
SubB-4	D′	A″	A′				91	97	6	Intermediate	Adjustment	0.24	0.45
SubC-1	E′	E′	E	7.3	5.5	−1.8	65	78	13	Visual	Intermediate	0.15	0.25
SubC-2	E	E	E				86	78	−8	Intermediate	Eidetic	0.39	0.26
SubC-3	C	C	C′				71	92	21	Intermediate	Adjustment	0.25	0.52
SubC-4	A′	A	D′				88	92	4	Visual	Intermediate	0.25	0.25

Table 7.
Comparison of learning effects among personality types, scores of pre-post tests, scores of first and second semesters information processing (cognitive) types, and correlation coefficient between response time to letters and reading time dividing into non-emotion or emotion contexts.

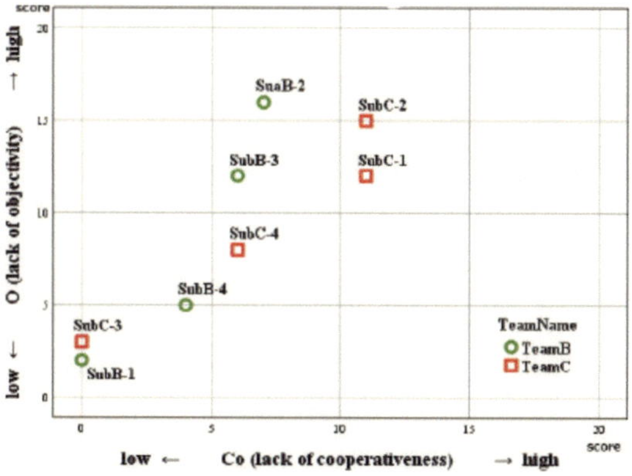

Figure 12.
Personality factor Co and O among twelve in YGPI.

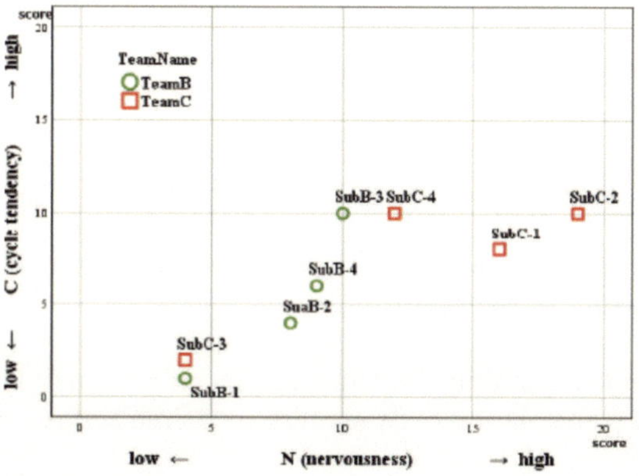

Figure 13.
Personality factor N and C among twelve in YGPI.

These results suggest strong correlations between lack of Cooperativeness and lack of Objectivity and Nervousness and Cycle Tendency. Moreover, a significant relationship was found between these personality factors and Cognitive Traits II, specifically those related to language information processing [12, 13, 37].

Based on the analysis of response times, the findings support the adequacy and reliability of these measures. They affirm the relevance of both Cognitive Style and Cognitive Traits, as individual characteristics of language information processing, in influencing outcomes in cooperative learning environments.

3.3 Indexical criterion value

Building on the previous section, it is proposed that three categories of individual characteristics (personality traits, Cognitive Traits I, and Cognitive Traits II) can serve as reliable indicators for measuring degrees of stress tolerance, specifically IIV-RT.

3.3.1 Features of basic types

a. Personality traits

A type intermediate between emotional stab and insecurity
B/E type emotional insecurity
C/D type emotional stable

b. Cognitive traits I (Visually, Auditory, Intermediate types)

Visual type:
In this type, there is no significant correlation between response time and reading time (with auditory presentation).
Individuals tend to process written language without phonetic encoding, resulting in shorter decision-making response times.
Auditory type:
In this type, a significant correlation exists between response time and reading time (with auditory presentation).
Individuals rely on phonetic encoding during written language processing, leading to longer response times for decision-making.

c. Cognitive traits II (Eidetic, Adjusting, intermediate types)

Eidetic type:
This type tends to process and make decisions independently of questionnaire content.
There is minimal correlation between response time and reading time in this type (with auditory presentation).
Adjusting type:
In this type, decision-making is influenced by questionnaire content.
There is a strong correlation between response time and reading time.

3.4 Comparison of traits among basic types

3.4.1 Characteristic traits

This section compares the basic types using the technostress tolerance level measurement developed by Kasuga [21].

- YGPI (Y-G Personality Inventory): This is a psychological assessment that uses questionnaires to measure stress tolerance.

- Kasuga's Questionnaire Method: This method is designed to assess degrees of stress tolerance.

Findings:

- D/C Types (YGPI):

  o High degree of stress tolerance

  o Emotionally stable traits

  o Individuals with high emotional stability tend to exhibit greater stress tolerance.

- E/B Types (YGPI):
  - o Low degree of stress tolerance
  - o Emotionally unstable traits
  - o Individuals with emotional instability tend to have lower stress tolerance.

Special care is required when reporting test results to individuals with emotional instability, as they may exhibit heightened sensitivity to feedback regarding stress tolerance.

3.4.2 Cognitive traits

- Response time shows the feature of information processing.
- The feature of information processing shows cognitive traits.
- Accordingly, Cognitive traits show response time.

3.4.2.1 Cognitive traits I (visually, auditory, intermediate types)

Cognitive Traits I can be identified by comparing response times between sound-based auditory response time (ARST) and letter-based visual response time (VRST) questionnaire presentations.

Individuals whose VRST is faster than the overall average tend to be of the Visual type. These individuals show low correlation between VRST and ARST. Visual type participants typically record shorter VRST and low correlation coefficients between modalities.

Individuals whose VRST is slower than the overall average tend to be of the Auditory type. These individuals show high correlation between VRST and ARST. Auditory type participants tend to exhibit longer VRST and strong correlation coefficients between modalities.

A learning effect was observed in the interaction between Visual and Auditory types. In the first semester, Visual Type participants had higher average learning outcomes than Auditory type participants. In the second semester, this trend reversed, with Auditory type participants outperforming Visual Type participants. Auditory type individuals are expected to show greater learning effects over time compared to Visual Type individuals.

3.4.2.2 Comparison of VRST between emotional and non-emotional questionnaire contents

The following features of information processing were observed:

- Participants whose correlation coefficient between reading time (sound voice) and VRST is higher for emotional content than for non-emotional content are classified as Adjusting Type.
- Participants whose correlation coefficient is lower for emotional content than for non-emotional content are classified as Eidetic Type.

- VRST for Eidetic Type participants tends to be highly variable across the questionnaire.
- VRST for Adjusting Type participants shows strong correlation, particularly in response to emotional content.

3.4.2.3 Relationship between personality traits factors and cognitive traits

Adjusting type individuals tend to receive lower scores on both the Non-Cooperativeness and Non-Objectiveness factors. They also tend to score lower on the Nervousness and Cycle Tendency factors. Individuals with low scores on Non-Cooperativeness and Non-Objectiveness also tend to score lower on Nervousness and Cycle Tendency.

Eidetic type individuals tend to receive higher scores on both non-cooperativeness and non-objectiveness factors. They also tend to score higher on the Nervousness and Cycle Tendency factors. Individuals with higher scores on non-cooperativeness and non-objectiveness also tend to score higher on Nervousness and Cycle Tendency.

According to the preceding analysis of individual cognitive traits based on response time, there seems to be strong alignment with the results of technostress tendencies measured via questionnaire [21]. This suggests that the results of Test 1 may serve as a valid criterion for measuring the degree of stress tolerance (IIV-RT).

3.5 Measurement method of stress tolerance: IIV-RT

Figure 14 shows the measurement instrument used to assess the degree of stress tolerance. This instrument investigates IIV-RT through a dual-task approach.

As shown in **Figure 15**, Mr. HARIS begins by explaining the instrument's operation, the structure of the dual tasks, and important reminders before the test begins. In addition to this oral guidance, Mr. HARIS asks participants a series of preliminary questions to confirm their identity, physical condition, hours of sleep, and so on. Participants respond verbally or via keyboard.

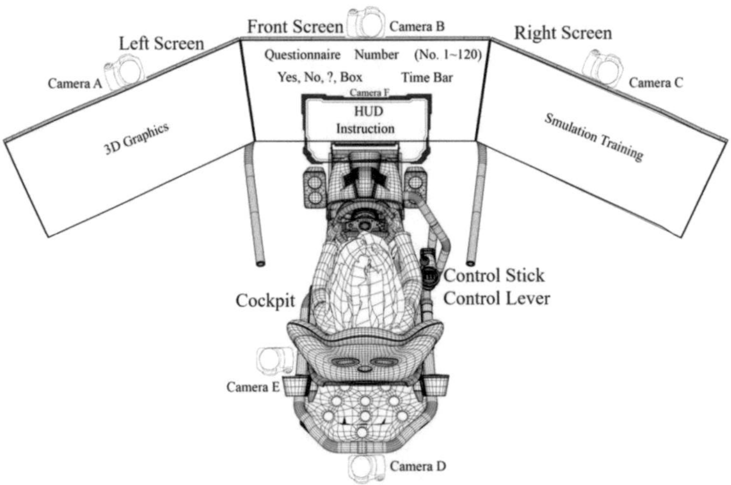

Figure 14.
The cockpit simulation, showing one of three detention types used for measuring stress tolerance via response and reaction time.

Figure 15.
Participants will be instructed by this AI robot, Mr. HARIS.

After the measurement is complete, Mr. HARIS provides participants with an explanation of their individual test results from Test 1 and Test 2. Lastly, participants are asked to complete a brief post-test questionnaire (**Figure 16**).

3.5.1 Measuring method of Test 2

3.5.1.1 Test 2-A firsthand task

Initially, the front screen (**Figures 14** and **16**), which is identical to the simulation training interface, displays only the questionnaire number. Immediately afterward, the questionnaire content is presented via sound voice (**Figure 17**, A-1). Participants listen to the audio through headphones (**Figure 17**, A-3). Once the audio presentation is done, three response buttons and a time indicator bar appear on the front screen (A-4, A-5; **Figure 17**). Participants select a response, either "Yes," "No," or "?," based on whether the desired behavior reflects their own daily life. The method of selection is shown on the Heads-Up Display (HUD) (**Figure 14**), and participants operate accordingly. When a participant makes a selection, the corresponding button is highlighted on the front screen (**Figure 16**). Simultaneously, the time indicator bar activates, beginning 2 seconds after the buttons appear. All indicators then fade out over the

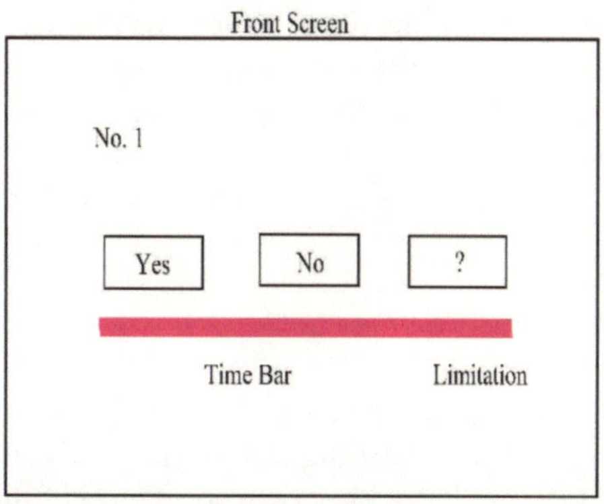

Figure 16.
Response box and timeline will be shown on the front screen.

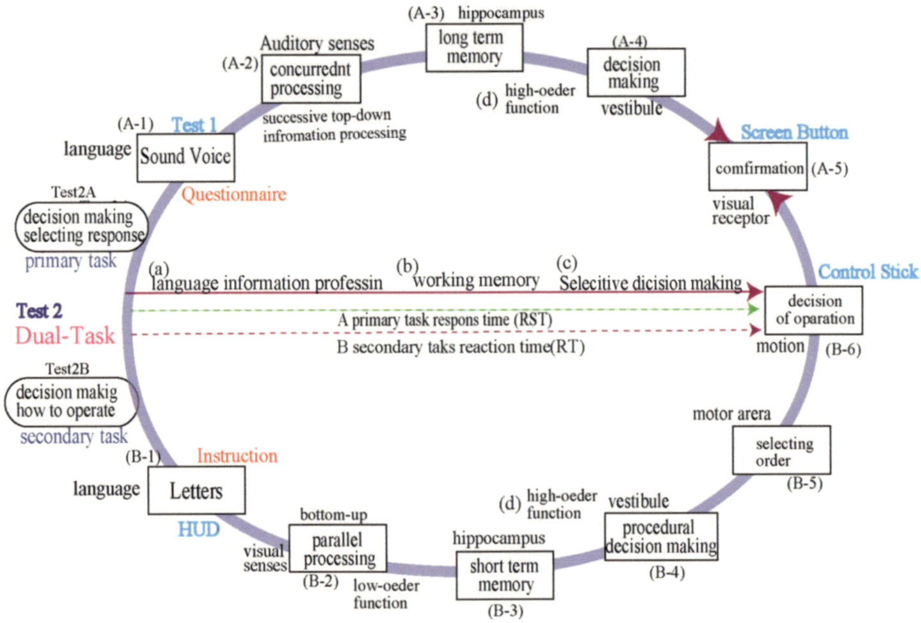

Figure 17.
A model of the dual-task method for measuring stress tolerance (IIV-RT).

course of 1 second. After the fade-out, the next questionnaire number is displayed on the screen. Response time is measured from the moment the questionnaire begins presenting to the moment the response button is highlighted (**Figure 17**, A-6).

3.5.1.2 Test 2-B secondary task

In front of the cockpit, a transparent HUD is projected onto the front screen (**Figure 14**). This HUD presents the operational instructions for selecting response buttons from five distinct sets (**Figure 17**, B-1). Each set contains three response options: "Yes," "No," and "?." These sets are cycled through every 24 items of the 120-questionnaire sequence (**Figure 17**, B-2). Upon completion of each 24-question block, a new checkbox is marked, accompanied by an engagement sound to signal the transition.

Participants are expected to follow the instructions displayed on the HUD (**Figure 17**, B-3) after determining their answer to the primary task (**Figure 17**, A-4,5). This process involves selecting the response that best reflects their awareness or behavior, as indicated in **Figure 4**, B-6 and B-7. Once a participant selects a button, it is highlighted on the front screen, allowing them to visually confirm their choice (**Figures 14** and **16**).

3.5.2 Measurement of motion reaction time in Test 2-B

When the number for the firsthand task is displayed, five depth cameras (**Figure 15**, Cameras A–D) begin recording the participant's movements. Each motion is tracked and recorded separately for every questionnaire item. The system captures full-body motion, and the duration from the initiation of movement to its completion is defined as the reaction time, as referenced in sources [40–42].

3.5.2.1 Method of data gathering for Test 2-B

Participants follow the operational instructions displayed on the HUD to select their response. After making a selection, they press the Enter key, which activates a pointer on the selected button on the front screen (**Figures 14** and **16**). The questionnaire is delivered via audio through headphones, and response buttons do not appear until the audio concludes. Because of this delay, participants need to remember their chosen response until they can confirm that the applicable bottom has been selected and highlighted. They are confirm that the appropriate button has been highlighted. They are allowed to change their within a 2-second window.

Once the response buttons appear on the screen, a time indicator bar begins progressing After 2 seconds, all visual indicators faded out over the course of 1 second. If a participant fails to complete the response within the timeframe, they are not allowed to repeat the questionnaire, unlike in Test 1, where they are permitted a second attempt.

Reaction time is defined as the interval from the appearance of the response buttons to the moment the selected button is highlighted. In addition, whole-body motions are recorded by six cameras, capturing the full duration of each action (**Figure 14**). This extended measurement is referred to as choice reaction time (CRT).

For each of the 120 questionnaire items, participants select "Yes," "No," or "?" based on whether the content reflects their own daily behavior. As the button positions remain constant on the front screen, participants may begin the selection process based on HUD instruction even before the buttons appear. Thus, CRT will be encompasses the entire motion sequence from the beginning of the questionnaire to the end of the session, including timeline records.

As this secondary task relies on auditory presentation, participants must engage in sequential memory processing aligned with top-down language comprehension. Although they cannot immediately select a button, they must wait for its appearance and subsequent highlighting to confirm the accuracy of their operation. This requirement places a greater cognitive load on memory in Test 2 compared to Test 1.

3.5.2.2 Measurement of RT and CRT by motions

1. Method of measurement and data recording

   Reaction time is measured based on motion events recorded by five depth cameras positioned around the participant during the task. Each motion event (M event) is tracked from its initiation to completion. For each event, three parameters are recorded: the body part involved (b), the spatial range and position of movement (r), and the time duration from start to end (t). These components are collectively referred to as Mbtr [43].

   By parallel processing, the system captures and records multiple data streams, including body parts ($b_1 \sim b_x$), time intervals ($t_1 \sim t_x$), and positional data ($r_1 \sim r_x$), for each M event. These datasets are sent to HARIS for further analysis. As outlined previously, the measurement procedures from Section 3.5.2.1 are repeated. The order and number of questionnaire presentations are also logged alongside Mbtr data to facilitate analysis of IIV-RT. Each session is sorted chronologically, from the moment the questionnaire number appears on the front screen to the point when the response buttons fade out. This timeline-based segmentation allows for the calculation of choice reaction time (CRT), using Mbtr data divided according to the appearance and disappearance of response buttons [44].

2. Response time

The selected response button and its corresponding event time are recorded on a timeline to confirm the questionnaire content of and its sequence ID. Response time is the interval between the appearance of the buttons and the moment the selected button is highlighted.

Data from Test 1 and Test 2 are processed using a concurrency control scheme. The analysis distinguishes between emotional and non-emotional content, comparing RT across intra-individual variables. Sentence length, which is measured by reading duration and character count, is included as a factor in the analysis.

In the case of Test 2, data processing must account for the dual-task structure. Reaction time is measured stereographically through motions, incorporating not only the timeline of motion initiation but also vector composition. This enables concurrent processing and comparative analysis across datasets.

The goal of measuring reaction time is to check each participant's level of stress tolerance. These results are communicated to participants to support self-monitoring and stress management. The cognitive load associated with working memory may increase as participants engage with the task, potentially enhancing performance across the 24-question blocks [45, 46]. As such, task difficulty may need to be adjusted based on individual stress tolerance levels. For these reasons, parallel processing and timeline-based comparisons across the 24 questionnaires are essential for meaningful analysis.

3.5.3 Method of evaluation for Test 2

The evaluation of a participant's current level of stress tolerance is based on the methods outlined in the previous section. It is expected that reaction time will improve relative to earlier performance. Reaction time is calculated using the formula:

$$RT = (b_1 \sim b_x) + (t_1 \sim t_x) + (r_1 \sim r_x) \tag{1}$$

To account for practice effects, alterations may be necessary, such as changing the order of questionnaire presentation or modifying operational procedures every 12 tasks. However, depending on individual differences among participants, the timing and nature of these changes will vary. Therefore, each participant's degree of stress tolerance should be predicted by both individual criteria from Test 1 and the analytical results from Test 2.

If a participant's reaction time increases compared to previous tasks, this may indicate a lower level of stress tolerance. In addition, direct effects, like the number of M events and the timing of those events along the timeline, may also need to be evaluated. These evaluation methods can be further refined by comparing results across different Test 1 types [47].

Participants may improve their stress management, control, and coping by receiving feedback on their evaluation results. However, the method of reporting these results should be tailored to individual differences. From this viewpoint, it is expected that HARIS will play a key role.

It is also possible to predict a participant's daily stress tolerance level using reaction time data from Test 2. In this case, it is considered that the way of reporting to participants should be differ depending on individual traits based on each criterion measure. Again, the reporting method should be adapted to each participant's traits, as defined by their criterion measures. This is based on the understanding that individuals experience and respond to stress in different ways, and that these responses are shaped by cognitive awareness.

Munakata's [48] framework suggests that stress arises through two pathways and defines stress function using three components: D, P, and S. According to this view, HARIS must exercise caution and sensitivity when providing individual support.

In the next section, we introduce a hypothesis-testing model to identify effective ways to help participants gain better control over their stress by understanding their individual patterns of information processing.

4. Hypothesis

4.1 Dual loop theory

As discussed in the previous section, this research investigates language information processing by using psychological questionnaires while also classifying individual cognitive traits into types. Based on these findings, as shown in **Figure 18**, Dual Loop Theory suggests that humans may process language information through two separate channels: positive feedback control (PFC) and negative feedback control (NFC).

For example, analysis indicates that individuals with the Adjusting Type cognitive trait may differentiate questionnaire content into emotional and non-emotional categories. Emotional content is likely processed by PFC, whereas non-emotional content may be handled by NFC.

Moreover, collaborative learning experiments with optimized groupings revealed distinct learning effects based on cognitive type. These findings correlate with earlier research on technostress tolerance [17, 21], which showed strong corrections between cognitive traits and personality factors.

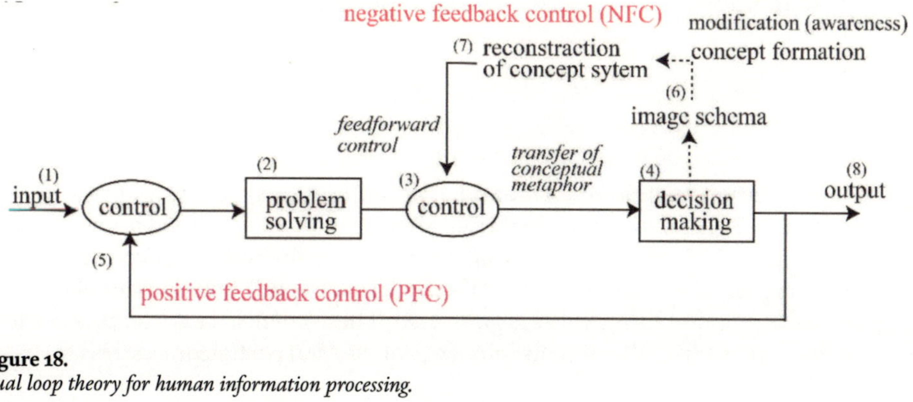

Figure 18.
Dual loop theory for human information processing.

Building on these insights, a new measurement method for stress tolerance called IIV-RT has been developed using a dual-task framework (**Figure 18**). Kasuga's studies reported that when participants received feedback on their technostress tolerance levels, their stress management performance improved. From this perspective, it is presumed that participants may enhance their stress control, coping, and regulation by receiving daily feedback on their stress tolerance through this method [17, 18, 43, 49].

4.2 Hypothesis model for measuring stress tolerance: IIV-RT

4.2.1 Language information processing

In the dual-task structure (**Figure 17**, Test 2A, Test 2B), visual instructions (**Figure 17**, B-1) are indicated by letters or codes on the HUD (**Figure 17**, B-2). These inputs are received through visual receptors and processed via bottom-up parallel mechanisms. Next, questionnaires are delivered via sound voice (**Figure 17**, A-2), and the language information (**Figure 17**, A-3) from auditory senses will be processed. In this case, it is considered that the language information from auditory perception is processed through top-down parallel mechanisms, reflecting sequential comprehension from auditory perception (**Figure 17**).

4.2.2 Memory

In the process of decision-making, participants compare the questionnaire content with their own daily life experiences, retrieving relevant information from their long-term episodic memory (**Figure 17**, A-4). After selecting a response (**Figure 17**, A-5) [40], this decision is stored in working memory. Participants must then recall the operational method for selecting the appropriate response button in the secondary task (**Figure 17**, B-4), enabling them to proceed with the dual-task response [41, 42].

4.2.3 Decision-making process

As shown in **Figure 17**, the selected response (**Figure 17**, A-5) informs the operational method (**Figure 17**, B-5), which is transmitted via the motor cortex to initiate the corresponding physical action (**Figure 17**, B-6). Once executed, the selected button is highlighted on the screen (**Figure 17**, A-6), allowing participants to visually confirm the accuracy of their response. This confirmation marks the completion of one cycle of the dual task (**Figure 17**).

The total reaction time is calculated using the formula:

$$RT = (b_1 \sim b_x) + (t_1 \sim t_x) + (r_1 \sim r_x) \tag{2}$$

During execution, it is assumed that the hippocampus, which is responsible for memory processing, experiences greater cognitive load than in Test 1. This increased load is expected to influence CRT, making it longer than the response time observed in Test 1. By aggregating CRT and reaction time across the full-body motion data, a comprehensive reaction time profile is generated. These results are analyzed according to individual criteria, allowing for observation of each participant's stress tolerance level (IIV-RT).

By reporting these results to participants regularly, it is anticipated that they will improve their self-regulation and develop more effective strategies for stress control, management, and coping.

4.3 Discussion

We answer the following research questions, which we presented in the Introduction, through the findings of this study:

1. Are there any relations between individual differences in language information processing traits [37] and stress?

   This question is clarified through the Dual Loop Theory (**Figure 18**) [24, 30], which explains how individuals, particularly those of the Adjusting Type, use different processing loops (PFC and NFC) depending on content types (**Figure 18**). These differences are related to technostress tolerance levels [21]. The dual-task method is expected to show both inter- and intra-individual variability in RT, offering deeper insight into cognitive processing under stress.

2. How should pilots control daily their own stress daily? Are there ways of measuring stress tolerance for self-management using predictable patterns?

   It is expected that participants can monitor their own intra-individual variability in response time by tracking total response time, the number of response changes, CRT, and so on. HARIS will provide personalized feedback based on these measurements, enabling participants to recognize and predict their daily stress patterns. This supports autonomous stress regulation.

3. What are the reasons for stress control using objective indicators?

   Stress perception varies widely among individuals. Kasuga's inventory for techno-stress tolerance revealed strong correlations with the personality factor Co (YGPI). These findings underscore the importance of self-observation and the strengthening of stress tolerance. Therefore, providing participants with objective criteria is essential for effective stress control and management.

5. Conclusion

This study proposed a novel measurement method for assessing degree of stress tolerance called IIV-RT as a preventive approach to flight accidents caused by SD. Drawing on prior research and the dual-task hypothesis model (**Figure 17**), the chapter outlines the conceptual and practical foundations of this method.

The implementation of a specific and effective measure was developed in response to reports indicating that novice pilots experience great stress in the cockpit environment. To simulate and reproduce these conditions, this method was integrated into a controlled simulation screen environment.

Conceptually, the approach involves measuring reaction time through full-body motion tracking while participants engage in dual tasks designed to investigate language information processing. By analyzing the total reaction time across these tasks, the system measures daily stress. Showing these results to participants is expected to help them control their stress autonomously.

In addition, the study emphasizes the importance of understanding language information processing within the measurement model. This relationship is essential to interpreting individual stress tolerance and developing strategies for stress control.

Author details

Keiko Tsujioka
Institute for Psychological Testing in Japan, Suita, Japan

*Address all correspondence to: tkeiko@sinri.jp

References

[1] Richard J, Stott R, Stott E. Orientation and disorientation in aviation. Physiology & Medicine. 2013;**2**:2

[2] Mizohata Y, Ohrui N, Kikukawa A, Kobayashi A, Fujita M, Takada K, et al. SpatialDisorientationandCountermeasure. JASDF. AML. Reports. 2016;**56**(4):79-93. ISSN 0023-2858, Aeromedical Laboratory, Japan

[3] Gibb R, Ercoline B, Scharff L. Spatial disorientation: Decades of pilot fatalities. Aviation, Space, and Environmental Medicine. 2011;**82**:1-8. DOI: 10.3357/AMHP.5446.2020

[4] Pennings HJM, Oprins EAPB, Wittenberg H, Houben MMJ, Groen EL. Spatial disorientation survey among military pilots. Aerospace Medicine and Human Performance. 2020;**91**(1):4-10

[5] Takada Y, Hisada T, Motohashi K, Kobayashi H, Naruo Kuwada N, Takeuchi Y. Subjective evaluation of a trial spatial disorientation profile. Reports Aeromedical Laboratory. 2009;**49**(4):59-67. ISSN 0023-2858

[6] Kilic B, Ucler C. Stress among ab-initio pilots: A model of contributing factors by AHP. Journal of Air Transport Management. 2019;**80**:101706. ISSN 0969-6997. DOI: 10.1016/j.jairtraman.2019.101706

[7] Ihara R, Kpga Y, Okazaki N. Support for student pilots to remain in the flight training course: An approach from self-regulated learning. JASDF. AML. Reports. 2021;**61**(1):1-11. ISSN 0023-2858, Aeromedical Laboratory, Japan

[8] Gillingham KK. A primer of vestibular function, spatial disorientation, and motion sickness. United States Air Force School of Aerospace Medicine. 1966:AD637943

[9] Osaka N. Awareness in higher brain function. Higher Brain Function Research. 2012;**32**(3):427-432

[10] Nilli L. Perceptual load as a necessary condition for selective attention. Journal of Experimental Psychology: Human Perception and Performance. 1995;**21**(3):451-468

[11] Lavie N, Beck DM, Konstantinou N. Blinded by the load: Attention, awareness and the role of perceptual load. Philosophical Transactions of the Royal Society. 2014;**B 369**:20130205. DOI: 10.1098/rstb.2013.0205

[12] Tsujioka K. Toward clarifying human information processing by analyzing big data: Making criteria for individual traits in digital society. In: Strategy and Behaviors in the Digital Economy. IntechOpen; 2020. pp. 125-140. DOI: 10.5772/intechopen.86037

[13] Tsujioka K. Toward clarifying human information processing: A case study of big data analysis in education. In: Karwowski W, Ahram T, editors. Intelligent Human Systems Integration 2019, AISC. Vol. 903. Switzerland: Springer Nature; 2019. pp. 378-383. DOI: 10.1007/978-3-030-11051-2_58

[14] Tsujioka K. A study of the impact of audio or visual media on decisions in Japanese: Comparing response and reaction times in tasks between emotionality and introversion-extroversion. In: Annuals of Educational Studies. Vol. 16. Osaka: Osaka University; 2011. pp. 33-44

[15] Wium J. Validating stress tolerance exercise with game-based assessment and

stress checklists. In: 21st International Symposium on Aviation Psychology. 2021. pp. 13-18. Available from: https://corescholar.libraries.wright.edu/isap_2021/3

[16] Bor R, Eriksen C, Oakes M, Scragg P. Pilot mental health assessment and support. A Practitioner's Guide, LCSH: Air Pilots—Mental Health, Air Pilots—Mental Health Services, Aviation Psychology. Routledge; 2017. LCCN 2016030855, ISBN 9781138222038 (hardback), ISBN 9781315401942 (e-book)

[17] Kasuga N, Takahashi A. Self-management of technostress: Using a questionnaire to evaluate technostress tendency. Japanese Society of Psychosomatic Medicine. 1996;**36**:483-488

[18] Kasuga N. Questionnaire to evaluate technostress tendency. Psychosomatic Medicine. 1994;**34**(400):414

[19] Oritsu M, Murakami M, Katsura T, Nozaki S. The reliability of the stress tolerance check list. Japanese Society of Psychosomatic Medicine. 1996;**36**(489):496

[20] Onaka T. Stress and its neural mechanisms. Folia Pharmacologica Japonica. 2005;**126**:170-173

[21] Kasuga N. Study on Technostress Syndrome (Report 2): Psychological characteristics which are associated with techno-stress tendency. Psychosomatic Medicine. 1992;**32**:391-398

[22] Marcel AJ. Conscious and unconscious perception: An approach to the relations between phenomenal experience and perceptual processes. Cognitive Psychology. 1983;**15**(2):238-300

[23] Tsujioka B, Sonohara T, Yatabe TA. Factorial study of the temperament of Japanese College Male students by the Yatabe-Guilford personality inventory. Psychologia. 1957;**1**(2):100-119

[24] Tsujioka K. A case study of using ICT in education with big data analysis of learners' traits: Toward clear information processing. In: Proceedings of the Human Interface Symposium 2018. Human Interface Society; 2018. pp. 589-594

[25] Tomar A, Polygalov D, Chattarji S, McHugh T. Stress enhances hippocampal neuronal synchrony and prolongs sharp-wave ripple. Journal of Experimental Psychology. Human Perception and Performance. 1992;**18**(2):471-484

[26] Tomar A, Polygalov D, Chattarji S, McHugh TJ. Stress enhances hippocampal neuronal synchrony and alters ripple-spike interaction. Neurobiology of Stress. 2021;**14**:100327

[27] Baddeley A, Allen R, Vargha-Khadem F. Is the hippocampus necessary for visual and verbal binding in working memory? Neuropsychologia. 2010;**48**:1089-1095

[28] Schouten JF, Bekker JA. Reaction time and accuracy. Acta Psychologica. 1967;**27**:143-153

[29] Toyokura M. Divided attention and dual task. Higher Brain Function Research. 2021;**41**(2):193-203

[30] Tsujioka K. Dual loop theory: Eidetic feedback control and predictive feedback control. In: Assistive and Rehabilitation Engineering. Vol. 4. IntechOpen; 2019. pp. 51-70. DOI: doi10.5772/intechopen.89681

[31] Randy G, Ercoline B, Scharff L. Spatial disorientation: Decades of pilot fatalities. Aviation, Space, and Environmental Medicine. 2011;**82**(7):717-724

[32] Kharoufah H, Murray J, Baxter G, Wild G. A review of human factors causations in commercial air transport accidents and incidents: From to 2000-2016. Progress in Aerospace Sciences. 2018;**99**:1-13. ISSN 0376-0421. DOI: 10.1016/j.paerosci.2018.03.002

[33] Munir A, Aved A, Blasch E. Situational awareness: Techniques, challenges, and prospects. AI. 2022;**3**:55-77. DOI: 10.3390/ai3010005. Available from: https://www.mdpi.com/journal/ai

[34] Selye H. The evolution of the stress concept. American Psychologist. 1973;**61**:692-699

[35] Cannon WB. Stress and strain of homeostasis. The American Journal of Medical Sciences. 1935;**189**:1-14

[36] Beste C, Yildiz A, Meissner TW, Wolf OT. Stress improves task processing efficiency in dual-tasks. Behavioural Brain Research. 2013;**252**:260-265. ISSN 0166-4328. DOI: 10.1016/j.bbr.2013.06.013. Available from: https://www.sciencedirect.com/science/article/pii/S0166432813003574

[37] Tsujioka K. Toward clarifying human information processing by analyzing big data: Comparing response time to questionnaires included heteronym word for problem solving between visually and Auditory types. In: Smart Trends in Computing and Communications: Proceedings of SmartCom 2020. Singapore: Springer Singapore; 2021. pp. 171-182. DOI: 10.1007/978-981-15-5224-3_16

[38] Tsujioka K. A case study of using big data processing in education: The method of matching members by optimizing collaborative learning environment. In: Social Media and Machine Learning. Vol. 3. IntechOpen; 2019. DOI: 10.5772/intechopen.85526

[39] Tsujioka K. Optimization for collaborative learning environments by matching team members with analyzing students' various data using ICT. In: IHIET 2020, SISC. Vol. 1152. Switzerland: Springer Nature; 2020. pp. 245-251

[40] Sternberg S. Two operations in character recognition: Ome evidence from reaction time measurements. Perception & Psychophysics. 1967;**2**:45-53

[41] Sternberg S. Memory scanning: Mental process revealed by reaction-time experiments. American Scientist. 1969;**57**:421-457

[42] Sternberg S. The discovery of processing stages: Extensions of Donders' method. Acta Psychologica. 1969;**30**:276-315. ISSN: 0001-6918

[43] Jensen AR. The importance of intraindividual variation in reaction time. Personality and Individual Differences. 1992;**13**:869-881

[44] Cannon WB. The mechanism of emotional disturbance of bodily functions. New England Journal of Medicine. 1928;**198**:877-884

[45] Baddeley AD, Hitch GJ. Developments in the concept of working memory. Neuropsychology. 1994;**8**(4):485-493

[46] Baddeley A. The episodic buffer: A new component of working memory? Trends in Cognitive Science. 2000;**4**:417-423

[47] Pitts RM. The information capacity of the human motor system in controlling the amplitude of movement. Journal of Experimental Psychology. 1954;**47**:381-391

[48] Munakata T. A theoretical overview of self-care & social support network,

perceived stressor and coping behavior-self-growth from "Iiko" behavior. Journal of Mental Health: Official Journal of the National Institute of Mental Health, NCNP, Japan. 1993;**6**(39):29-40

[49] MacDonald SWS, Nyberg L, Backman L. Intra-individual variability in behavior: Links to brain structure, neurotransmission and neuronal activity. Trends in Neurosciences. 2006;**29**(8):474-480